Perspectives on Development in the Middle East and North Africa (MENA) Region

Series editor

Almas Heshmati, Sogang University, Seoul, Korea (Republic of)

This book series publishes monographs and edited volumes devoted to studies on the political, economic and social developments of the Middle East and North Africa (MENA). Volumes cover in-depth analyses of individual countries, regions, cases and comparative studies, and they include both a specific and a general focus on the latest advances of the various aspects of development. It provides a platform for researchers globally to carry out rigorous economic, social and political analyses, to promote, share, and discuss current quantitative and analytical work on issues, findings and perspectives in various areas of economics and development of the MENA region. Perspectives on Development in the Middle East and North Africa (MENA) Region allows for a deeper appreciation of the various past, present, and future issues around MENA's development with high quality, peer reviewed contributions. The topics may include, but not limited to: economics and business, natural resources, governance, politics, security and international relations, gender, culture, religion and society, economics and social development, reconstruction, and Jewish, Islamic, Arab, Iranian, Israeli, Kurdish and Turkish studies. Volumes published in the series will be important reading offering an original approach along theoretical lines supported empirically for researchers and students, as well as consultants and policy makers, interested in the development of the MENA region.

More information about this series at http://www.springer.com/series/13870

Leila Benali

Electricity-sector Reforms in the MENA Region

Evaluation and Prospects

Springer

Leila Benali
Institut d'Etudes Politiques
Paris, France

ISSN 2520-1239 ISSN 2520-1247 (electronic)
Perspectives on Development in the Middle East and North Africa (MENA) Region
ISBN 978-3-030-07168-4 ISBN 978-3-319-96268-9 (eBook)
https://doi.org/10.1007/978-3-319-96268-9

Softcover re-print of the Hardcover 1st edition 2019

This Springer imprint is published by the registered company Springer Nature Switzerland AG
The registered company address is: Gewerbestrasse 11, 6330 Cham, Switzerland

"Rien n'est possible sans les hommes, rien n'est durable sans les institutions"
"Nothing is possible without men; nothing is lasting without institutions"
Jean Monnet

Preface

This book coincides with interesting times in the global economic history and in the history of the electricity sector in particular. Unprecedented changes in the global economic and energy landscapes are happening. The world tries to emerge from the profound recession of 2009, which followed several years of growth and an unprecedented escalation of oil prices and capital costs. After waves of privatizations and deregulations in several parts of the world, the last decade witnessed episodes of utilities reregulation.

Energy in general and electricity in particular remains a magnet for political intervention. This is not just another study advocating hybrid models as a convenient solution for countries searching for excuses because deregulation is challenging technically and politically. Trade-offs have to be made between pragmatism and theory. This analysis takes into account practical models of deregulation.

The electricity sector is perceived as an essential public good, ubiquitous and unavoidable. All consumers are direct or indirect stakeholders, including in developing countries where access to the service, sometimes at low fee, is a priority. To add to the complexity, electricity cannot be economically stored or transported over long distances with the current and foreseeable State of technology, contrary to oil and gas. Therefore, it might be one of the rare "commodities" to really justify the creation of regional markets. Moreover, as with commodities, limited supply leaves room for market manipulation, if the market is weakly regulated.

Capital intensity calls for the creation of mechanisms to increase investors' confidence (sovereign guarantees, long-term commitments and take-or-pay clauses). It is usually the State which serves as the backbone of these mechanisms. Last but not least, there could be significant demand and supply adjustment lags if price was left to do all the economic work, and because supply interruptions (and price volatility) would be unacceptable to governments, the latter would opt for the least complex capacity mechanism of all: direct intervention. Public authorities would then have all the good reasons in the world to remain involved in the electricity sector, particularly in immature, rapidly growing markets.

This research explains the reforms' dynamics in the MENA region by evaluating possible organizational configurations in the case of the ubiquitous and strategic electricity sector, with a focus on the period between 2000 and 2008. This period was indeed the most fertile in terms of efforts for electric sector reforms and coincided with escalating oil prices, in addition to profound changes in the region, at the macroeconomic, social and political levels. Very limited electricity sector reforms occurred after 2008 until today, given the spillover effect of the global financial crisis on the MENA economies followed by the waves of social unrest of the so-called Arab Spring.

In order to rationalize the differences between the dynamics of reforms in industrialized countries as opposed to those in developing nations and then to the MENA countries, the hypotheses of some of the principal trends of the institutional economy, particularly that of Douglass Cecil North, will be used. The reforms' efficiency is "measured" with theoretical evaluation criteria, using an ad hoc grid. The criteria of (1) attractiveness, (2) feasibility developed by the World Bank (1995) and Menard and Shirley (2000), (3) credibility of these reforms (Levy and Spiller 1994/1996) and (4) the remediability test of Williamson are used. The latter will evaluate the reforms' capability to keep open the possibility of enhancement alternatives. Another goal is to complement existing analyses of the single buyer model (SBM) and electric reforms in developing countries. One alternative which can address these countries' needs without jeopardizing the gains already achieved and without adding tremendous political and financial reform costs is proposed.

The contribution of this research to this body of neo-institutional evaluation techniques is twofold. First, this evaluation grid is applied to the MENA reforms for the first time. Also, the novelty is to combine it with two additional evaluations: with a quantitative evaluation (based on prices and investment incentives) and with an analysis of the sequencing of reforms and power market development. Second, this approach to electricity reforms' evaluation is complemented by using the general equilibrium theory and welfare economics. This second methodological choice stemmed from the observation that successful reforms are usually part of a comprehensive framework within an economy-wide reorientation for a sustainable development path. The current thinking on the creation of sustainable wealth has not developed any analysis of the characteristics of hydrocarbonexporting countries for instance. One key question is whether current fuel pricing policies are sustainable. Fuel pricing policies have been at the core of the long-term fuel supply agreement "paradigm" which served as one of the region's main competitive advantage to attract Independent Power Projects (IPPs) and energy-intensive industries. An underlying question is whether these countries are becoming genuinely wealthier or impoverished as a result of their fuel pricing policies, with a focus on two empirical examples in the region: Saudi Arabia, as an oil exporter/heavily populated country, and Qatar as a gas exporter/lightly populated country.

The first two chapters are historical. They provide the elements to understand the factors that led these countries to question the organizational models of their electric systems. The historical perspective is essential to highlight some specific characteristics of the reforms' dynamic in the region during the different phases of recent history. The characteristics have a direct impact on the institutional dynamics and the strategic choices. Moreover, a comprehensive historical analysis allows us to "deconstruct" preconceived notions which are widely endorsed by the specialized literature and the (rare) analyses of electricity reforms in emerging countries.

Chapter 1 examines the impact of the unique confluence of demographic, social, economic and political development in the region. Electric capacity and investments needed in the region are substantial. In economies and demographics in transition, assessing the needs cannot be done in isolation from prudent macroeconomic assumptions on development needs.

The MENA region is unique, with respect to its hydrocarbons' resources endowment, its demographic trends, the rapid urbanization and industrialization it witnessed, its spectacular economic growth and massive investments that accompanied it. Most resource-rich countries have explored ways to monetize their energy resources and sought to create more jobs domestically. Some countries embarked on ambitious industrial plans supported by below-market energy prices. In parallel, electricity residential tariffs have been widely subsidized in most countries in the region. Resource-poor countries benefited directly or indirectly from the economic booms of their neighbours.

Therefore, it is not surprising to find in this region the world's highest electricity demand growth rates and the highest consumption per capita. Obviously, there is a wide discrepancy between highly populated North African countries, hydrocarbon-rich gulf countries and fertile east Mediterranean countries. However, the social challenges, the urgent need to create employment and the need for infrastructure, housing and various utilities are common.

Most energy demand models have underestimated the energy needs of the region. In 2008, the general consensus estimated that approximately 210 GW needed to be installed between 2006 and 2015 in the region, at a cost of around $20 billion (bn) per annum. The Arab Spring and the various wars that have been impeding the development of the region create substantial needs of reconstruction or delayed investments. APICORP for instance estimate that in the period 2016–2020, the region will need to actually invest $334 bn in its power sector. Of this, $198 bn will be needed to add 147 GW of generating capacity, while the rest should be invested in transmission and distribution (T&D). But for the reasons stated earlier, these "econometric" needs represent a fraction of the actual demand needed for the populations to prosper and the economies to flourish.

There are alternatives to building new power plants, particularly when decision-makers believe in the cyclical nature of costs and can delay the investment decision. These alternatives include demand-side techniques and supply-side strategies (efficiency). They allow for limited capacity gains in comparison to a large power plant and assume a certain level of sophistication of the market and additional investments to be spent. However, it is important for all countries to implement efficiency programmes and techniques to take advantage of the "hidden generation capacity". More importantly, these programmes will contribute to the development of a sound regulatory framework, of a higher knowledge of the electric system, and will send the right signal to the customer on the value of energy.

These alternatives will not be sufficient to cover the large needs of these countries. In this context, similar to other regions, the growth in electricity demand, the sense of urgency and the inability of the governments to respond efficiently all played a major role in urging these countries to restructure their electricity sectors.

Chapter 2 analyses the main factors which spurred reforms in the region. In addition to the classical drivers often associated with reform movements (poor performance of the State-owned/run electricity sector, inability of the sector to self-finance expenditures, etc.), there are three main dynamics which appear dominant: the need to capture revenue from privatization of assets, the urgency to increase the performance of the sector and the appeal to align a call for political reform with some form of economic reform. Chapter 2 questions the traditional paradigm stating that electricity reforms in the region can only be analysed through the "oil prism": the introduction of Independent Power Projects/Producers (IPPs) in the region has often been linked to the fluctuation of oil prices and the State of the governments' budgets. This was probably true in the early 1990s, but this general statement proved wrong several times in the post-2000 period. During the last decade, with the recent period of high oil prices, Middle Eastern countries benefited from an aberration: high oil prices without stagflation in the economies of the major consumers of their hydrocarbons, and in their own economies.

The experiences of reform of the different countries in the region are very enlightening in this regard. The behaviour(s) of the State, as a Group of Economic Orientation, is a critical driving force. Conventional wisdom would assume that countries would seek to capture a new form of rent. Oilexporting countries would be particularly impelled to do so in an environment of low oil prices and decreasing oil revenue. Non-oil exporting countries would be tempted by these "exceptional receipts" from asset sales to fill in budget deficits. However, even those who focused their budgetary policies on privatization revenues quickly understood that privatizing assets without making the effort of restructuring them might not be as lucrative as originally thought. Moreover, there is a limit to what a country can privatize in quantity and in scope. Finally, the increasing needs of the population for basic utilities including water and power spurred governments to rationalize the management of their budget. In parallel to this rationalizing mood, the role of the State has also been evolving. National budgets started to be preferably allocated to activities that are *more naturally* within the responsibility of the State, and this conditional allocation assumes an implicit ranking of the State's tasks, in order to define which ones will be "privatized" first.

In some cases, revenues from privatizations might have lured some governments and encouraged them to open their electric sectors to foreign and private capital. When that was the case, the reform occurred amid a wave of privatizations impacting several sectors of the economy. However, in the long run, the behaviour of the State vis-à-vis privatization, as a preamble to electricity reform, becomes more dictated by strategic economic orientations rather than opportunistic short-term budgetary considerations.

On performance, accelerated investments and high demand growth rates displayed inefficiencies in all parts of the sector. Conversely, it appears that major blackouts cannot be necessarily explained by the perceived weaknesses of monopoly structures (general lack of reliability, cost reduction mechanism, efficiency incentive, etc.).

Finally, in some cases, electricity reforms were driven by a general movement of economic reform, which was in turn *just substituting to a call for political reform.* Then, the decision-makers face several dilemmas. The reform is expected to continue to serve the interests of the population, even if utilities are expected to be treated as commercial commodities, rather than social benefits. In very few instances, energy and electricity reforms in the region were the heirs of defence policies, with the primary objective to attract foreign interests in the region to guarantee its defence. Therefore, the close involvement of the highest levels of the State, in addition to the interior or defence minister sometimes, can facilitate electricity reform programmes but can also complicate them.

The third dilemma facing the governments in the region is the need to diversify the economy because of the depletion of fossil fuels, or their scarcity. Decision-makers were hoping to achieve efficiency and fuel gains by introducing the private sector in utilities.

Therefore, up until the middle of the last decade, most countries in the region were actively restructuring their power and water sectors for various reasons. By 2006, there were only four countries which were completely closed to Independent Power Projects for various, mainly political, reasons: Libya, Syria, Kuwait and Iraq. Today, with the exception of Kuwait, any major electricity reform is bound to fail in the existing fragile political and industrial set-up of the three other countries.

Between 2008 and 2010, reforms slowed down and the few moves towards more deregulation were generally shelved. The deceleration of deregulation occurred in oil producing and non-oil producing countries alike. This coincided with the "reregulation" process in already deregulated markets in other parts of the world. MENA countries were discovering the limits of full deregulation at the same time as the rest of the world.

As a matter of fact, post-2010 and until today, the single buyer model (SBM) remained as the dominant market structure in the region and all countries which were considering to move beyond the SBM either froze or reversed their reform plans. In 2017, it could be argued that the only exception could well be Saudi Arabia with its long delayed reform. The National Transformation Plan (NTP) unveiled in 2016 includes a target for electricity generated through strategic partners to increase from 27 to 100% by 2020. This was meant to accelerate the unbundling and accelerated transformation of the national utility, before privatizing the four generation companies that were going to be created out of it. In reality, the Saudi electricity sector is only pursuing its long overdue plan of sector restructuring including unbundling which has been on the drawing board since the early 2000s. The proposed model is not very different from the evolution proposed in this book.

Finally, as regional interconnections are usually credited for their potential to save future capacity investments, their relation with reform programmes should not be overlooked. With a group of State-owned utilities, the nature of the institutional set-up in each country would not act as an impediment for cross-border interconnections. As the different experiences in the region showed, the usual obstacle is funding and lack of common vision. In fact, the current electricity interconnections have benefited from the fact that electricity has been perceived as the "easy way out" for regional integration. Other integration mechanisms (common currencies, economic convergence or even gas trade) have faced more serious obstacles. Having said that, electric reforms in a specific country could be challenging for the regional vision.

Chapters 3, 4 and 5 build an ad hoc evaluation tool in order to evaluate electricity sector reforms in MENA. In this evaluation, two steps are suggested.

In Chap. 3, a more "common" method which consists in assessing the impact on production costs (performance) and on investments (incentive) is used. It is important to take into account the timing and sequencing of reforms in the region. In all cases, the path chosen led to various degrees of market power that could prevent further reforms.

Today, economics cannot be narrowly defined as a social science concerned with the production and allocation of material goods anymore. Rather, it is about incentives motivating societal agents in social institutions. It is the spirit of this general statement that guided the evaluation grid in this analysis. It could be applied to other social, macroeconomic or political reforms. It could particularly be used when benefits and losses are difficult to quantify. Evaluating electricity reforms in developing countries suffers from lack of transparent data. The absence of retail price signals, because of implicit and unclear subsidies, clouds the picture, particularly if wholesale costs are not accessible. Until recently, demand response has been often overlooked and marginalized in the analysis. The difficulties in assessing the impact on production costs, efficiency and performance are in itself a major conclusion for the evaluation. Regulators have a difficult task. Only a handful of countries started recently to communicate on costs of electricity and subsidy levels, pressed by energy pricing reform requirements.

In the absence of a transparent and direct link between generation costs and retail prices, the impact of electric reforms in the MENA region on costs and investments is analysed. The impact on costs is unclear, but at least, one conclusion is that electric reform did not lead to a substantial decrease in costs even if expectations of economic performance played a critical role in launching these reforms in the first place. The RPI-X regulation, dominant in the region, should be acting as an efficiency incentive for most generators and that government-owned assets will be gradually in competition with the others.

The impact on investments is more clear-cut: electric reforms contributed to attract some investments in the region but failed to cover all investment needs. For instance, the "fuel offer", embedded in long-term "$1-gas" supply contracts, made the region attractive for power developers, but it was not sufficient, and more importantly, it was not sustainable. Future quantitative evaluations will be even more difficult as fuel prices are being reviewed and inflation of capital costs was passed through the system. The role of institutions which can be expected to perform this evaluation exercise, ideally independent regulators, becomes even more critical.

That would assume, however, access to relevant data and information and availability of resources to conduct the evaluation.

In Chaps. 4 to 8, an evaluation tool is used to evaluate electric reforms in MENA. This tool is based on a grid of four conceptual principles: attractiveness, feasibility, credibility and remediability (or the capacity of the reforms to keep possible future alternatives of enhancement). In MENA, countries reformed their electricity sectors differently. Even if all went for the single buyer model, they chose different paths to reach it. The paths are diverse in the number of steps taken, the length of the process and the sequence of the steps. Long reforms were not always the most successful, but quick reforms bear the risk of being defective. Long reforms can be interpreted as the sign of rigid industrial structures and complex institutional set-ups and might not solve the issue of market power. The region still displays an important concentration of players in the generation of electricity. Highly concentrated markets can have some advantages (similar to the advantages offered by monopolies) in some cases if they are properly regulated, but their sustainability can be questioned. To be sure, the attractiveness, the feasibility, the credibility and the remediability of these reforms can be used as qualitative evaluation metrics.

The concept of *attractiveness* (economic), with desirability (political), is central to the analysis of reforms. The attractiveness and desirability of a reform are usually driven by three main factors: an economic crisis, a sectorial crisis or a major political change. However, this theory, relevant in the case of reform vanguards like the UK, disregards the impact of "exogenous factors" and does not appear relevant in the case of several countries in the Middle East and North Africa. All reformers in the region were not driven by a sectorial crisis or by a major political change. In some countries, economic crisis might have probably been the most relevant factors in the pre-2000 period, but surely not afterwards. The history of electricity reforms in the region shows that attractiveness and desirability were mainly driven by other factors which do not fall necessarily in the three above-mentioned categories. The focus has tended to be on the introduction of competition, and exogenous factors (international trends, historical factors, etc.) were important in that process. The impact on macroeconomic policies (sought industrialization, economic diversification) is a key component of the attractiveness of MENA reforms.

The *institutional feasibility* relates to the degree of centralization of powers and decision-making. The industrial feasibility measures how easily transformable the industrial structure could be. The more decentralized the institutional environment is, the less feasible the reform will be without negotiation and spending of political capital. The more rigid the industrial structure is, the longer the reform will be. Time is a critical factor as it increases the probability for the reform to be distorted, reversed or frozen. However, transaction costs would be higher in countries with a rigid industrial structure, and it would be more difficult to reverse or to freeze a reform in an advanced stage.

In a region where world-class oil and gas reserves are concentrated, a feasibility analysis would not have been complete if it did not look at how the resource endowment of a country affects the feasibility of a reform and its process. The case of Saudi Arabia is particularly unique in the way its resource endowment has been influencing the reform path of its electricity sector.

Chapters 6 and 7 focus on the credibility and remediability of these reforms. The *credibility* of the reform—or its capacity to provide stability of commitments, and the ability of its governance structure to solve regulatory problems between the government, the operators and the other stakeholders—is critically supported by the design of "effective" regulatory institutions, meeting a set of criteria. In MENA, clarity of roles, participation of relevant stakeholders and accountability are usually guaranteed. The transparency of the decision-making process is often fulfilled by economic regulation and open bidding procedures, except under specific circumstances. The autonomy from political intervention and the predictability of decisions are less certain. More concretely, the design of effective regulatory institutions hinges on the availability and the use of an adequate supply of economic resources, particularly of human resources. Most gulf countries have been able to attract high calibre international economists in their regulatory authorities. Other countries in the region, particularly in North Africa, struggle to attract the few human resources they had, because of the "brain drain" they are suffering from in most sectors of the economy. The development of robust regulatory frameworks and strong institutions was hampered by under-funding and a reluctance of authorities to transfer independent decision-making to regulatory authorities (even when required to do so by law). Transitional arrangements are needed (initial price controls in contracts or licences or clear tariff setting principles in legislation), but enhancing regulation should be a continuous and evolving process. The reality is that this is easier in some countries in the region than in others.

Some countries (Morocco, Lebanon and Bahrain) decided to engage in a reform process without the creation of a regulation authority. This option bears cost and risk advantages for the State, but in general, an effective regulation in the absence of a regulatory authority assumes that the controlling role of local/regional authorities and professional associations is reinforced. The regulation would be non-discretionary, usually ex post (while a regulatory authority would regulate ex ante), less information-intensive, discontinuous and dependent vis-à-vis the political framework. In the context of vertical integration, network managers tend to deploy opportunistic behaviours to indirectly limit competition.

Regional experiences show that the coexistence of a strong State and a dependent (sometimes weak or absent) regulator did not prevent the move to the single buyer model (SBM) and the introduction of competition. It sometimes expedited the process. In the future, governments might be less willing to give sovereign guarantees if competition for projects increases, and if some competitors can find other ways to mitigate the risks without transferring them entirely to the State. Successful IPPs in the region were built on the credibility of the entities, on the predictability of decisions and on the clarity of the processes.

The credibility of these reforms involves analysing the subsidies issue and pricing of energy (electricity and fuel, as in several MENA countries, the electricity sector is subsidized by applying a "double subsidy"). When competition is introduced and electricity demand is growing, the growth in the volume of subsidies is usually not offset with the hypothetical decrease in production costs due to efficiency and competition. A subsidy applied through fuel pricing can potentially be counterproductive to stimulating efficient use of energy and lowering generation costs. Increasing fuel prices would encourage generators to invest in energy-efficient technologies. It would also increase the cost of electricity, hence increasing the level of direct subsidies to the electricity sector, assuming the same retail price. A simulation model enabled to compare wholesale prices with theoretical generation costs under different assumptions for project locations and fuel costs. The restriction of fuel supplies obliged most countries to rethink their position vis-à-vis subsidies. Regionally, the value of natural gas is becoming increasingly linked to, at least, its cost of supply or to its export netback value. A few countries have started to negotiate more sophisticated terms (gas prices indexed to end-user pricing or to alternatives) in recent contracts with some industries. But there are still several constraints on the way to review subsidies which transcends the political difficulty: macroeconomic policies, lack of alternatives, etc. In order to achieve a minimum level of credibility, authorities are encouraged at least to have a clear idea about their energy subsidies (quantify them) and to make their amount (including implicit subsidies) and the calculation methodology transparent. A few countries have started doing that (Egypt, Abu Dhabi, Oman and more recently Saudi Arabia); others will have to follow suit should they want to guarantee that minimum level of credibility for their electric reforms.

Then, as the ideal model, which could be duplicated to several countries, does not exist, it appears important to maintain some level of flexibility in order to allow improvement. The test of *remediability* of Williamson enabled us to conclude that, in the 12 countries which reformed their electricity sector, the single buyer model (SBM) appears to be a relevant model so far. The analysis of the advantages and shortcomings of the SBM as it applies to the region shows that its limitations are more related to the institutional and macroeconomic environment in which it is deployed than with its intrinsic weaknesses. But this is not an excuse to maintain a long-lasting status quo. The growth in consumption centres, creating large enough distribution companies enabling smart consumption, and the creation of large industrial centres, with a potential interest—cost or volume—to sign bilateral contracts with generators or trade electricity in wholesale markets, are a reality. Therefore, there are several areas in the region where an "SBM plus" could be one *feasible alternative*. In this work, the SBM plus is defined as an SBM model where bilateral contracts could be signed. The conditions to implement these contracts efficiently include the existence of more than one creditworthy entity to keep government away from contract and to give customers choice, access to transmission, robust system and demand management. It is interesting to highlight that, as the writing of this book is coming to an end, Saudi Arabia, the only country in the region effectively considering deregulation, is moving forward with a transformation of its electricity sector from the single buyer model to a "principal" buyer model, something similar to the SBM plus model suggested earlier.

However, market power could be a major impediment to remediability. In that sense, most electric reforms in the region do not fulfil Williamson's remediability criteria. Even in those countries which tried to break up monopolies and shift some power away from incumbent, the temptation to create (or recreate) national champions is great. Moreover, breaking up monopolies without giving the new entities the full means to develop is obviously counterproductive. If a national champion is needed or desired, it is preferred to keep politics away from it. However, the need to maintain the "sacred" consumer protection should play a balancing role and guarantee remediability of the reforms. Consumer protection is probably as important as power protection given the implicit "social contract" between the authorities and the population. Moreover, industrialization policies are built upon the notion of protection of industrial customers—against abusive behaviour concerning fuel or power supply.

Chapters 9 and 10 test the future of the reforms in the region against a set of policy decisions and orientations that could be, arbitrarily or not, chosen by governments. Indeed, the region is facing new challenges, issues and opportunities which could have an indirect impact on reforms. The two most pressing and critical sets of policy orientations are the electricity fuel mix options and resources monetization. The focus will be on hydrocarbon exporting countries in the region. The core question is whether these countries are building wealth or becoming impoverished over time, given the energy and fuel pricing policies they are following to support their industrialization strategies, their economic growth and their welfare principles. GDP or GDP per capita as metrics for sustainable wealth have shown their limits. The application of a theoretical framework to two major hydrocarbon exporting countries of the region, Saudi Arabia and Qatar, turns up surprising results. For example, current domestic pricing policies do not necessarily lead to a long-term impoverishment, but nothing proves that this policy is optimal. Saudi Arabia, a middle-income country according to the World Bank classification, appears to be building wealth more rapidly than Qatar, a high-income country.

Another underlying goal of this chapter is to revisit the determinism of the Dutch disease on MENA oil exporting countries and the general assumption of low economic performance associated with oil producing countries. Indeed, contrary to historical trends, some MENA hydrocarbon producers appear to be building genuine wealth, but probably not the way one would think.

Out of the 18 countries in MENA, 12 are hydrocarbon exporters (eight of which are OPEC countries). Hydrocarbons dominate the fuel mix, and natural gas, barring a few exceptions, has long been considered as the fuel of choice for power generation. The rich gas reserves position of the MENA region should support that choice. However, it is the deliverability of those reserves which questions the ability of indigenous natural gas to cover all electricity generation needs. The analysis concludes that with current rates of production and development of reserves, natural gas will need to be complemented by other fuel sources. In some countries, oil-fired generation continues to prevail because of unavailability of the gas or because of strategic reasons, but other alternatives are considered.

The main alternatives to hydrocarbons in electricity generation in the region are renewables, nuclear and coal. Another alternative source of supply is the "hidden" generation or demand management. All these complementing alternatives pose institutional, financial and practical issues.

With gas prices maintained at around $1–1.5 per MMbtu, other alternatives cannot compete on a pure cost basis, but gas prices have been slowly changing. Moreover, with the exception of coal, other alternatives will have to be directly or indirectly supported by the State, as a hedge against uncertain hydrocarbon supplies or missed CO_2 "opportunity charges". Nuclear carries strategic weight, but its development on a massive scale in the region would require heavy institutional, regulatory, educational and diplomatic investments and will take a long time to materialize. Even in situations where hydrocarbon reserves and supplies are abundant, at a stable cost, subsidies growth, because of growing demand, would lead to a reconsideration of fuel pricing policies. The reality is that, ultimately, the price will have to evolve, as marginal fuel production costs are increasingly higher than fuel prices agreed under several Fuel Supply Agreements in the region. Evolving fuel prices could be viewed, at first sight, as challenging for any introduction of an additional degree of deregulation, but there are various steps, including a period of assessment and testing, which should be followed before tariffs change. And in realty, there is an evolution as gas prices for utilities have been increased to higher levels, between $1.25 and $3 per MMbtu. New fuel pricing is applied in varying degrees, depending on the emphasis and the impact sought. It could be applied to support efficiency-related initiatives, to discriminate between end-use sectors or to encourage fuel displacement. The transition is meant to accompany the country as it transforms from a fuel exporter to a fuel monetizer and then to a mature economy.

Electric reforms are not in contradiction with a fuel diversification policy, but introducing new fuels and new technologies would require the State to step in directly or indirectly—through the upstream sector or the investment arms. This is also relevant for green initiatives and renewable, as having a centralized policy planning and a high institutional feasibility should, in theory, be an enabler. Regulation will have to play a critical role to ensure that the electricity cost structures and the efficiency of the system are maintained or enhanced and that the right investment signals are sent at the right time. Changes in fuel offerings (fuel type, pricing, contract terms, etc.) require a delicate balancing act. They are testing already established market models in the region and will challenge decisions to deregulate further.

The other major question is whether electric reforms are robust in view of resource monetization and sustainable development needs. In this regard, most MENA countries failed to make genuine investments in the 1970s and 1980s during periods of high oil revenue. Today, this is changing and it is increasingly admitted that natural resources depletion should be offset by manufactured and human capital increase. But there are still wide discrepancies between countries, particularly between North African and Gulf hydrocarbon exporters. The interpretation is that the latter would have benefited from a "mutualization of interests", while Algeria and Libya endured, among other things, the existence of vested interests which had time to organize and to benefit from subsidies with no pressure to seek efficiency or genuine investment.

It is possible to assess whether countries in MENA are building wealth or becoming impoverished over time, whether their hydrocarbons are properly monetized or whether the depletion of their natural resources is not offset by genuine economic investments. Pricing resources below their "social cost" often leads to inefficient consumption. Others would argue that this underpricing contributed to industrial and private sector development. In today's conditions, long-term fuel supply agreements with below-market prices are questioned and sometimes renegotiated. Contributions of the various "benefiters" to the economy are scrutinized and evaluated.

The application of the theoretical framework to two major hydrocarbon exporting countries of the region, Saudi Arabia and Qatar, turns up surprising results. Saudi Arabia, a middle-income country according to the World Bank classification, appears to be building wealth more rapidly than Qatar, a high-income country. Understanding all the caveats inherent to the calculation is important. For example, Qatar being more vulnerable to the pricing of gas than Saudi Arabia is to oil, there is a risk of understating or overstating wealth creation in Qatar. For all the uncertainties in the calculation, the two countries seem to be improving the way they manage their economies and the way they price and manage their resources. The last statement is particularly true for Saudi Arabia who exports its light crude, consumes its heavy crude and discounted products and sells its gas domestically at below-market price.

The pricing focus could initially be designed to support efficiency-related decisions—combined cycle, grid reinforcement and integration, efficiency programmes in industries. Using the fuel supply cost as a reference will be helpful in this regard, without being too punitive for the economy. Indeed, below a certain ceiling, fuel price increase (to electricity generators and industrials) will have a gradual impact on energy intensity without impacting GDP. If more price impact or fuel displacement is needed, pricing could rise to support the next expensive option (fuel alternative or market price). In this case, GDP will be directly affected. For example, export-oriented industries might relocate production to other countries. Conversely, that would free up more hydrocarbons for exports, potentially increasing the government's direct revenues. However, in hydrocarbon producing countries where fuel supplies are becoming tight, a redistribution of GDP might be necessary. Instead of energy-intensive industries, economy might be reoriented towards less energy-intensive services. For example, Dubai and Bahrain, which were suffering from dwindling oil output, have tried to turn their countries into financial hubs for the region. Today, a significant part of these economies is based on the financial-related sector. The downside is that this makes them more vulnerable to a global economic slowdown than their neighbours. The macroeconomic reorientation could, however, be long and costly, even painful. Decreasing drastically the energy intensity of an economy opens economic, political and, more importantly, social issues.

Research has been developed on the link between energy prices and energy intensity. Most analyses have been done ex ante and are based on econometric models. A developing body of research is assessing the impact of price changes on energy intensity, on GDP and then on wealth creation and genuine investment rates. Simulations could guide future policy decisions regarding investments, macroeconomic orientations and fuel pricing. There are increasing calls to differentiate fuel price by end-user type, e.g. the industries and the electricity sector, depending on the goal pursued.

Chapters 10 and 11 reemphasizes that the ideal model, duplicable to several countries, does not exist. At the same time, the State identity would evolve: from a Rentier State to a Developer State and from a fuel exporter to a fuel monetizer. The evaluation concludes that the SBM was so far adequate in most MENA countries. As a guide for the future, the SBM plus, allowing for bilateral agreements between suppliers and eligible consumers, was built so that it fulfils most of the four qualitative criteria used in this evaluation.

The SBM plus is attractive since it builds on the advantages of the SBM. It is feasible because it does not call for major additional institutional or industrial changes. It is credible in the sense that it reinforces a centralization of planning and a decentralization of risks. It is also credible since it would facilitate fuel and energy pricing revisions (particularly if contracts are shorter). Finally, policy-makers should ensure that enough flexibility is embedded in the model to ensure its remediability, despite the strong temptation to create, recreate or strengthen a national champion, and for the government to continue to be involved in the sector and its pricing administration. At the time of writing, at least one country in the region is seriously considering this model.

Paris, France Leila Benali

Acknowledgements

This work sheds light on the wider reforms' dynamics in the MENA region by evaluating possible organizational configurations in the case of the ubiquitous and strategic electricity sector, with a focus on the period between 2000 and 2008. This period was undoubtedly the most fertile in terms of efforts for electric sector reforms in the region. The thinking evolved over many years, but the main conclusions are standing the test of time, particularly during this era of unprecedented volatility in energy markets, and profound changes in the region, at the macroeconomic, social and political levels.

I would therefore like to thank H.E. Abdullah al Shehri (ECRA), M. Driss Benhima (RAM), M. Jean Paul Bouttes (EDF), Pr. Claude Henry and Pr. Jacques Le cacheux who reviewed an early version of this manuscript. I am particularly grateful to Pr. Jean Paul Fitoussi, M. Hubert Deslongchamps (Total) and Pr. Jean-Marie Chevalier for their trust and guidance during the early phases of this research. The list of colleagues, mentors and associates to thank in the wider energy industry is too long. Some, in oil and gas companies, banks, utilities or regulatory authorities, within and outside of MENA, have generously shared their perspectives on the energy dynamics in the MENA region. Over many years, H.E. Dr. Majid Moneef has been an excellent sounding board and a constant source of encouragement. Others, particularly at Cambridge Energy Research Associates (CERA), now IHS Markit, taught me precious lessons of resilience in analysis, in strategic thinking and in life: Dr. Daniel Yergin, Mrs. Vera de Ladoucette, M. Paul Markwell and others. Colleagues at Saudi Aramco provided much-needed peer review: M. Anwar Hussein, formerly at the UK's Department of Energy and Climate Change (DECC), M. Komborero Shoko and M. Alseni Bah. Ms. Amal Benali helped in the final editing of the book.

Finally, this book would not have materialized without the continuous and unconditional support of my family, my parents and my partner in life and particularly of my beloved Sarah and Nael who were patient enough to stand the multiple absences of their mother.

Abstract

This book offers an evaluation of MENA electric reforms and suggests an evolution based on this unique region's constraints and opportunities. Most countries face difficult trade-offs, while new challenges and opportunities emerge. But the dilemma remains: how to weigh short-term tangible costs against long-term hypothetical benefits?

The first chapter presents a historical perspective, an analysis of the factors behind these reforms and an assessment of capacity and investment needs. The relationship between these reforms and each of oil prices and regional interconnections is examined under a new angle.

Then, a quantitative and qualitative (attractiveness, feasibility, credibility and remediability) evaluation of these reforms is proposed. The single buyer model appears suitable, albeit not optimal. The proposed evolution (SBM plus) has to be feasible and flexible in order to allow improvement. It will have institutional, technical and regulatory implications. The identity of the State evolves in line with its involvement in the electricity sector. The region's new national champions would be dual firms, politically well-connected but economically pragmatic. Consumer protection and political structures should balance market power. Regulation is strongly needed.

These reforms' future is tested in view of two policy considerations: the fuel mix question and the sustainable wealth creation. Surprisingly, fuel pricing policies do not necessarily lead to a long-term impoverishment, but this path might not be optimal and fuel pricing has to evolve anyway. New fuel pricing are applied in steps, directly impacting electric and industrial sectors, as the country transforms from a fuel exporter to a fuel monetizer. Electric reforms do not contradict fuel diversification policies, but introducing new technologies requires the State to step in again. Economic, environmental and social sustainability cannot entirely be left to market forces.

Keywords Electric reform · Single buyer model · Genuine wealth creation · Regulation · Fuel pricing · Dual firms · Reforms' evaluation

Contents

List of Figures

List of Tables

Chapter 1
Introduction

Abstract The end of the 1990s witnessed the questioning of the classical organization of the electricity sectors of the Middle East and North Africa (MENA). This book comes at an interesting time. The world is emerging from years of economic growth followed by a major recession. In parallel, oil prices and capital costs had escalated dramatically. Consumers' bills (in deregulated markets) and governments' budgets (in highly regulated markets) were put under increasing pressure. Government intervention has been more dynamic. The need for such a study stemmed from three main reasons: a clear absence of a complete and comprehensive study on electric reforms in the MENA region; the unique concentration of hydrocarbon reserves in the region; the relative success of the MENA region as a market for investment in the electricity sector. This is a region with remaining substantial needs over the coming two decades at least.

The electricity sector has unique characteristics which influence the sector's structure. It is a capital-intensive industry where balancing supply and demand is challenging. For a nation, the electric system is a vital infrastructure. Electricity is a commodity that cannot be economically stored with current technologies, and electricity grids are complex and fragile. The analysis is built on the evaluation of existing structures and the measurement of the feasibility, attractiveness, credibility as well as remediability of current and future reforms. Moreover, because the electricity sector cannot be analysed in isolation from the rest of the economy, such an evaluation exercise would not be complete without a more holistic approach, integrating macroeconomic policies and development needs.

One first emerging conclusion concerns the central role in the region of the single buyer model (SBM), or wholesale competition with a unique buyer. The region today is facing new challenges and issues which in turn call for new policy orientations. It is timely for some MENA countries to consider *feasible alternatives* to enhance their sectors' structures, with the caveat that market power could have already significantly developed to make future reforms less feasible. In parallel, some MENA countries are making a better use of their resources and their revenues than during former periods of high oil prices. It will be important to shed light on whether the fuel pricing policies in the region and the way the electricity sector is treated enables to build wealth and

L. Benali, *Electricity-sector Reforms in the MENA Region*, Perspectives on Development in the Middle East and North Africa (MENA) Region,
https://doi.org/10.1007/978-3-319-96268-9_1

sustainable growth. In other words, the determinism of the *Dutch Disease* as it applies to hydrocarbons' exporters in MENA is questionned.

The end of the 1990s witnessed the questioning of the classical organization of the electricity sectors of the Middle East and North Africa (MENA). This phenomenon differs greatly from the dynamic of contestation of monopolistic structures which occurred during the 1970s in the western world. It is worth noting that this dynamic itself differed widely in the MENA region from one country to another.

During 2009, the world economy contracted, the financial crisis tightened lending and governments moved back to regulate the economies and, by extension, the electricity business in many regions of the world. In parallel, oil prices and capital costs had escalated dramatically. Consumers' bills (in deregulated markets) and governments' budgets (in highly regulated markets) were put under increasing pressure, including in the MENA region, where most countries benefited one way or the other from increasing oil and gas prices. In this context, government intervention has been more dynamic, particularly given the high capital expenditure needed in the power sector. Globally, re-regulation has occurred in some markets to help stabilize electricity tariffs, while other countries started considering market-oriented reforms to their energy pricing. Locally in MENA, government involvement and electricity reforms are being reshaped.

The need for a comprehensive analysis of electric reform in MENA stemmed from three main reasons:

- *First, there is a clear absence of a complete and comprehensive study on electric reforms in the MENA region*, despite the existence of a wide array of analyses of electricity reforms applied to different regions in the world. This observation is reinforced by the review of existing literature (Jamasb et al. 2004; World Bank 2005) which concludes more generally that "rigorous empirical and policy-relevant evidence on the performance and determinants of electricity reform in [developing] countries is rather limited".
- *Second, these countries are "unique" because of the concentration of hydrocarbon reserves in the region.* Each of them is either an important oil or gas producer, or is located next to a critical player in the energy scene. This concentration will not only impact the fuel choices in the energy mix but also the policy orientations of the governments and the strategies of the different players.
- *Finally, the MENA region as a market for investment in the electricity sector has witnessed some success during the last decade, although this success was limited in comparison to other regions like Latin America.* The booming periods of investments depended on a large set of factors, discussed in the first chapters. This is a region with a high growth potential (7–10% and even beyond in some countries) and large investment needs during the next two decades.

The analysis of the electricity sector reform is an excellent opportunity to shed light on the profound socio-political changes happening in the MENA region and the economic reforms attempted by different stakeholders. Indeed, the electricity sector is perceived as an essential public good, ubiquitous and unavoidable. All consumers are direct or indirect stakeholders, including in developing countries where access to the service, sometimes at low fee, is a priority. This analysis will go beyond the "classic" framework of evaluation of sectorial reforms. Because the electricity sector cannot be analysed in isolation from the rest of the economy, such an evaluation exercise would not be complete without at least a more holistic approach which integrates macro-economic policies and development needs.

In order to rationalize the differences between the dynamics of reforms in industrialized countries as opposed to those in developing nations and then to the MENA countries, the hypotheses of two principal trends of the institutional economy are used. The first is that of D. North, which insists on the necessity to take into account other factors than the simple "calculating rationality", such as ideologies. The second is that of Williamson, which insists on the "atmosphere" surrounding the reforms, either creating, encouraging or hampering them. North's contribution to the Institutional Economy puts the role of institutions at the nexus of economic development and reforms' analysis. Institutions are defined as the formal and informal rules of behaviour that govern economic transactions. Formal rules are the laws and regulations enforced by the government and the informal ones refer to rules of behaviour enforced by peers. In general, societies are governed by both these formal and informal rules, but the informal rules are more important because they represent the actual implementation of formal rules.

The electricity sector has unique characteristics which influence the sector's structure. For a nation, the electric system is a vital infrastructure. It is a highly capital-intensive industry where balancing supply and demand is challenging. Electricity is a commodity that can not be economically stored with the current technologies, and electricity grids are complex and fragile. The analysis is built on the evaluation of existing structures and the measurement of the feasibility, attractiveness, credibility as well as remediability of current and future reforms. This theoretical framework is applied to empirical experiences in the region. These experiences vary from vertically integrated monopolies (Libya, Kuwait), single buyer models (UAE and Morocco), mixed models (post-reform Morocco, Israel) or models where monopolies and local IPPs coexist (Iran). Actually, there are various ways to view the business model of the electricity sector, beyond the usual four organizational models presented in Appendix B: institutional differentiations, distribution of overseeing and regulation tasks or the type of industry design are other ways to define the business model in a specific country. The industry design could be ownership based (state-owned or controlled, privately owned), function-based (vertically integrated, unbundled) or pricing mechanism-based (wholesale or retail competition). A reform would seek to move from one model or design to the other.

1.1 The Starting Point and the Questioning

The starting point of this work lies in an observation: electric systems in the region do not automatically fulfil the favourable conditions to deregulation presented by Joskow (2000): (a) Demand stagnation and chronic overcapacities, as the sector gets mature; (b) an advanced stage in production, transport and information flow management techniques' evolution. However, these countries had to face the imperatives of markets' liberalization, sometimes pushed by the requirements of the international financial institutions. The international climate during the 1980s and most of the 1990s was filled with pro-privatisation and pro-market ideologies, helped with the discourses of the World Bank and the IMF.[1] It was «dans l'air du temps».[2]

In parallel, demand for additional generating capacity and transmission infrastructure has been rapidly strengthening, mainly due to demographic growth and pursued industrialization. It implied huge financing needs, difficult to drain from the national budgets. It was important to identify: when to proceed? How much is needed? And who would provide the finance to serve these needs? Countries had to be able to reform their electric industries and to find ways to finance these reforms. Of course, the need for a market-oriented reform and the urgency of this need varied widely between the countries across the region, but the race for sources of financing was real, for hydrocarbons' exporting countries and the others alike.

Countries in MENA with a rapidly growing domestic demand for power face two primary sets of challenges: developing power generation projects to meet this demand and establishing an efficient and well-designed power industry. If the only goal was to satisfy the electricity demand regardless of the efficiency of the sector, a capacity building program via investment in power plants would be sufficient, but the issue is much more complex. Also, if competition *has* to be introduced, it is critical to define *where* this competition is actually possible (Littlechild 1998; Percebois and Wright 2001; Hunt 2002) along the chain. A clear distinction between a negotiated introduction of reform and an imposed one needs to be made. The «bet of Joskow & Schmalensee»[3] has been often used to encourage deregulation, however, most countries in the region have introduced competition in generation only.

---

[1]The World Bank and the IMF were created following the financial devaluations of the 1930s and the willingness to create a stable financial system and an international commercial system. Both promoted controlled liberalism. With the rise of market-oriented ideologies in the 1970s, they encouraged a more intense liberalization by imposing conditions related to external opening and the reform of state-owned firms. The Asian crisis, the failure of public aid to development and the bankruptcy of US hedge fund LTCM, questioned that rationale and resulted in a higher political intervention to find ad-hoc solutions for each nation. See Béatrice Hibou, *Economie politique du discours de la Banque Mondiale en Afrique sub-saharienne, du catéchisme économique au fait (et méfait) missionnaire*, les Etudes du CERI, no. 39, mars 1998.

[2]M. Jean-Paul Villain, First Advisor, Investment Strategy, Abu Dhabi Investment Authority (ADIA), interview of May 22 2002.

[3]Joskow and Schmalensee (1983) bet that the reform can only be a positive sum game, i.e. that transaction costs incurred because of the reform will be offset by production and efficiency gains.

Was it economically attractive or politically desirable? This region illustrates perfectly the difference between the two concepts of attractiveness (economic) and desirability (political) of a reform. In this context, the parameters governing the industrial organization of the different actors in the sector, the institutional set up and the issue of regulation are explored.

1.2 The Theoretical Framework

First, the analytical framework uses theoretical tools for the evaluation of existing structures. The two factors in favour of the integrated monopoly model, according to Finon (1997) and Hunt (2002), can be summarized in two categories of coordination prerequisites: technical coordination and transactional coordination. Conversely, integrated monopolies usually suffer from a lack of efficiency incentives and problems of technological adaptability (Joskow 1989, 1991); therefore, they rarely offer endogenous mechanisms to reduce production costs. This general statement is reaffirmed in Dumez and Jeunemaitre's (1991, 1998, 2000) research on factors which explain the lack of efficiency incentives in the European air traffic services sector. These factors include the absence of a set of mechanisms: market pricing, competition and monopoly regulation. Here, the same factors will be used to test empirical experiences of monopolies—or quasi monopolies—in the regional electric sector.

Second, this research explains the reforms' dynamics by analysing and evaluating all possible organizational configurations. *The reforms efficiency is "measured" with theoretical evaluation criteria, more specifically with an ad-hoc grid re-built for this exercise.* Perez (2002) was the first to apply such a framework to evaluate electric reforms in Europe. The analysis presented will call upon the criteria of ***(a) attractiveness***, ***(b) feasibility*** developed in World Bank (1995) and Menard and Shirley (2000) and ***(c) credibility*** of these reforms (Levy and Spiller 1994/1996). Furthermore, it appears important to use the ***(d) remediability*** test of Williamson, in order to appraise the reforms' capability to keep open the possibility of enhancement alternatives. *A complementary theoretical goal is to suggest limitations and practical additions to existing analyses of the Single Buyer Model and electric reforms in developing countries.* Potential alternatives are analysed through the spectrum of possible public-private partnerships and possible contractual relationships between the various players in the electricity sector. This in turn creates different regulatory requirements, and various interactions between the regulator and the different actors, including the government. *One alternative in particular can address these countries' needs without jeopardizing the gains already achieved and without adding tremendous political and financial reform costs.* Williamson's methodology uses transaction costs as a parameter to select governance structures in the relations between the economic actors. But more importantly, *the resource endowment of the region and the underlying energy subsidies are characteristics that will be included in the qualitative evaluation. The addition to the mainstream evaluation* *techniques will mainly consist in applying this grid of qualitative criteria to MENA reforms and in combining it, first, with a quantitative evaluation (based on price and investments' incentives), and, second, with an analysis of reform sequencing and market power.*

Third, the case for having a more comprehensive assessment of these electric reforms was presented. In order to offer a holistic approach to electricity reforms' evaluation, another major branch of economics, the general equilibrium theory and welfare economics, is used. First, most major MENA countries are in economic transition and the electric reforms in the region are primarily driven by macroeconomic development needs. Second, this evaluation exercise needs to confront a set of policy decisions and orientations (including welfare management) that could be, arbitrarily or not, chosen by the governments. Third, current research on genuine wealth creation still lacks a robust assessment of hydrocarbons' exporting countries. Here, this work attempts to bring a contribution to this body of research. The main question is whether the current fuel pricing policy to the electricity sector (and to industries) is sustainable. Are these countries becoming genuinely wealthier or impoverished as a result of their fuel pricing policies? This book focuses more particularly on two empirical examples in the region: Saudi Arabia, as an oil exporter/heavily-populated country, and Qatar as a gas exporter/poorly populated country. Fuel pricing policies were at the core of the long-term fuel supply agreement "paradigm" which served as the region's main competitive advantage to attract Independent Power Projects (IPPs) and energy-intensive industries. In both countries, these policies are evolving.

1.3 The Methodology

A historical perspective, to understand what led these countries to question the organizational models of their electric systems, appears essential and necessary for two main reasons:

- *To highlight some specific characteristics of the reforms' dynamic in the region, specifities that have a direct impact on the institutional dynamics and the strategic choices*: the countries' trajectories (demographic, industrial, economic, urban), despite being diverse within the region, are quite unique in the world. Since the early 2000s, the new dynamic of the governments' investments, locally and forcefully, with less "recycling of petrodollars@" abroad, has completely changed the region's face in comparison to where it was just a decade earlier. Furthermore, the exceptional concentration of oil and gas resources obviously impacts the fuel mix decisions. More interestingly, this concentration changes completely the role of the electricity sector. It elevates it beyond its mere utility function as a critical driver of economic activity, beyond an already strategic status. For some hydrocarbons exporters, electricity could become an important part of the hydrocarbons exporting strategy (see Chap. 10), and it is influenced by these same strategies. This impacts the governments' energy policies and the behaviours of the different players involved. This is not to mention the sector's central role in rent-management or incentives distribution.

- *To "deconstruct" preconceived notions widely endorsed by the specialized literature and the (somewhat rare) analyses of electric reforms in emerging countries.*[4] A historical analysis coupled with this deconstruction, largely used in social sciences, will allow to confirm or refute some postulates, with the goal of building a more rigorous framework for the analysis. For example, in the first chapter it is demonstrated that IPPs' introduction, and more generally the choice of reform, is not (or is not anymore) necessarily linked to oil price movements and to the state of governments' budgets. Periods of low oil prices, creating important budget deficits for oil exporting countries, do not necessarily push countries to encourage privatization, as is commonly stated. Likewise, the opposite is challenged: that periods of high oil prices lead automatically to the freezing of reform programs. The usually cited drivers such as needs for financing and economic performance are also questioned. These drivers, even if they are not necessarily reached ex-post, partially explain the dynamic of reform ex-ante, but that they are not the only ones. The mutations of the role of the State, its *liberal interventionism*, the regional (cross-countries electric interconnections, integration policies) and external dynamics are additional important factors.

Policy objectives and economic challenges, driven by a unique confluence of demographic, social, economic, political development create additional institutional requirements, which in turn tend to "distort" the reform's process. In addition to testing the limits and successes of the reforms, these different forms of "policy distortion" are examined. The future amount of capacity and investments needed in the region are substantial. As regional interconnections are usually credited for their potential to save future capacity investments, their relation with reform programs could not be overlooked.

In completely deregulated markets, price plays a central role and has to do all the economic work: dispatch and allocate resources, chase market power, send the signal at the right moment for new supply and reduce consumption when shortages are threatening. Whether it is actually able to perform its "economic duties" is another debate. In less deregulated markets, the transition from a monopoly structure to any other structure often raises the same questions about the benefit of the reform, particularly as it usually comes at a cost (institutional redesign, political cost, contractual costs). However, it is generally difficult to assess that benefit, even ex-post. This evaluation suggests two steps. First, it uses a more "common" method which consists in assessing the impact on production costs (performance) and on investments (incentive). Lack of availability and inconsistency of data across the region makes this exercise challenging. Moreover, the timing of reforms coincided

---

[4]A review of existing literature (Jamasb et al. 2004; World Bank 2005) shows that "rigorous empirical and policy-relevant evidence on the performance and determinants of electricity reform in (developing) countries is rather limited".

with major changes in the international and regional industrial landscape, with consolidations and costs escalations to cite a few. The evaluation based on comparative analysis in a changing environment would have limits. Therefore, it appears critical to scrutinize the timing and the sequencing of the reforms in the region. The diversity of paths can bring additional insight to the evaluation exercise, even if there is no ideal path. In all cases, whichever path was chosen would have led to various degrees of market power that could prevent further reforms. Then, an evaluation tool is used to test these reforms. This tool is based on a grid of four conceptual criteria: *attractiveness, feasibility, credibility and remediability* (or the capacity of the reforms to keep possible future alternatives of enhancement).[5] The application of this framework will take advantage of the diversity in the region in terms of supporting examples and experiences.

The region is facing new challenges, issues and opportunities which could have an indirect impact on reforms. The policies adopted by the governments behave like exogenous factors, vis à vis the electricity sector. They could influence reform, but they are exogenous in the sense that they could be pursued regardless of the electric reforms. The two most pressing and critical sets of policy orientations are the electricity fuel mix options and the resources monetization.

The diversification of the fuel mix is becoming high on governments' agendas for different reasons (environment, strained fuel supplies, geopolitics). This poses the important question of fuel and electricity pricing. In parallel, countries are increasingly concerned about the need to achieve a "sustainable" growth in their development, while optimizing the use of their natural resources. *The core question is whether these countries are building wealth or becoming impoverished over time, given the energy and fuel pricing policies they are following to support their industrialization strategies, their economic growth and their welfare principles.*

The last two chapters consist in testing the robustness of these reforms against the future and in light of these policy orientations. The decision to test the robustness was comforted by the context of global economic downturn and high volatility of oil prices post-2008.

1.4 Emerging Conclusions

An appropriate institutional framework to deliver the reform and to monitor the market in the long run is necessary, but perhaps not sufficient for the reform to be "successful". This hypothesis will guide this work and will encourage to combine a neo-institutional evaluation with a macro-economic analysis. One first emerging conclusion concerns the central role in the region of the single buyer model (SBM), or wholesale competition with a unique buyer. It appears that most MENA countries

[5]Concepts developed respectively by World Bank, Menard and Shirley, Levy and Spiller, and Williamson.

think they would not get more benefits out of a "next step", whatever that next step would be, at least not until the SBM is properly set up and fully functioning. Even if the implementation of the SBM in MENA countries has been imperfect (like in other parts of the world), it looks like the model could remain relevant for the foreseeable future. However, the region is facing today new challenges and issues which in turn call for new policy orientations. It may be about time to think about the next step but it is also critical for countries to identify their needs to go beyond the SBM.

Obviously, general challenges inherent to electric systems remain relevant in MENA, in addition to the region-specific issues. The independence of regulators in relation with other institutions needs to be reinforced at various levels. A new market structure would require an increasing need for coordination. That could be even more challenging in MENA given the existing coordination issues in some electric systems. The various contractual agreements (power and water purchasing, fuel supply) will need to be self-consistent and also consistent with the prevailing legislation. In the interest of all parties involved, these agreements also need to be sustainable and robust enough to undergo future policy evolutions or political and economic conditions. Finally, the SBM could create, recreate or strengthen, powerful companies which could be outside of political control.

Performance, efficiency and costs decrease were supposed to be some of the primary goals of introducing competition. However, the result of reform is never guaranteed ex-ante, and this is the biggest dilemma of policymakers: how to make acceptable short-term costs that might (or might not) bring long-term gains? Actually, international experiences showed that deregulation does not always lead to lower wholesale or end-users tariffs. Other regions' experiences led sometimes to increase in prices and/or limited incentives for investment.[6] The political capital spent would be perceived to be higher than the hypothetical gain of a more advanced reform process, with a more deregulated market. Therefore, it might look more suitable to also justify reforms on the basis of shorter term qualitative benefits (better transparency, better view on demand, accountability).

In any case, it appears difficult to maintain the status quo in an environment of emerging new challenges and opportunities (tight fuel supplies, climate change or local pollution, tough investment climate, regional power trade, development needs). Moreover, *there are additional looming challenges often overlooked and that would question the sustainability of the electricity market designs adopted in the region.* The first is the need to encourage investments in networks' adequacy and robustness while the focus has hitherto been on generation. Moreover, recent developments and opportunities offered by distributed energy solutions and smart grids would need to be leveraged. The second relates to the potential for demand growth deceleration in

---

[6]For example, in the UK, for a typical annual household electricity consumption of 3300 kWh, the cost was £62 in 2001, just before market opening. It fluctuated between £57 and £62 until 2006 when it jumped to £88. Source: British Gas, UK's Department for Business, Enterprise and Regulatory Reform.

the region. Current models do not address these two aspects, future arrangements and market structure will need to take them into account.

For all these reasons, it may be about time for some MENA countries to consider *feasible alternatives* to enhance their sectors' structures (an SBM-plus alternative), with the caveat that market power could have already significantly developed to make future reforms less feasible, as highlighted in this book. This forms the basis of the second main emerging conclusion.

The electricity sector cannot be designed in isolation from the rest of the energy sector and the rest of the economy. The traditional linkages between the different fuels and energy forms are being reconsidered. Macro-economic considerations and development needs are core. But the resource endowment of the region is hard to ignore. Pragmatism will be needed to design, implement and manage increasingly large integrated projects in an integrated and yet-to-be properly designed institutional framework. Some MENA countries are making a better use of their resources and their revenues than during former periods of high oil prices. Their resource allocation policy prioritizes between exports and domestic demand and between various segments of domestic demand. Fuel pricing is used as a means to achieve this prioritization. This work will shed light on whether the fuel pricing policies in the region and the electricity sector's organisation enable to build wealth and promote sustainable growth. It will question the determinism of the *Dutch Disease* as it applies to hydrocarbons' exporters in MENA.

Chapter 2
A Unique Confluence: Demographics, Socio-Economics and Politics

"A long-run view of history reminds us of the presence of changes, ruptures, and discontinuities. It should warn us against simply extrapolating from a brief period of a few years, and projecting the future simply as a continuation of the immediately lived and experienced past."
—Professor Harold James, Princeton University and author of The End of Globalization: Lessons from the Great Depression

Abstract The organization of the sector in the different countries of the MENA region has been changing overtime. These countries share several common features which impacted the governance and the organisation of the electricity sector, but can widely differ in terms of economic and human development standards, sources of revenue, private sector development, credit ratings, to cite a few. The social challenges, the urgent need to create employment, the need for infrastructure, housing and various utilities are common. This region is unique and accounts for the world's highest electricity demand growth rates and the highest consumption per capita. High demographic growth, rapid urbanisation and accelerated industrialization are some of the first limits faced when attempting to apply generic models. Economic growth is "distorted" by international oil prices. The presentation of the first three experiences of electric reforms in the region (Morocco, UAE and Oman) provides interesting insights on the peculiarities of this region.

Keywords Investments needs · Modelling · Drivers of growth · Distortion

The organization of the sector in the different countries of the MENA region has been changing over time, following a series of historical, economic and political changes. The history of electricity in the region dates back to the end of the nineteenth century. Egypt is the first country in the region and—one of the first in the world—to introduce electricity in cities (Cairo). It developed an electricity

L. Benali, *Electricity-sector Reforms in the MENA Region*, Perspectives on Development in the Middle East and North Africa (MENA) Region,
https://doi.org/10.1007/978-3-319-96268-9_2

industry, with heavy involvement of private utilities. The situation changed with the advent of socialism in the fifties and the wave of nationalizations in the sixties.

All the countries in the region witnessed, after World War II and the end of their respective periods of colonial rule, a fairly moderate growth of their electricity sector. Barring some exceptions (Israel), rapid growth of the electricity industry did not happen in the region before the nineties. As in other parts of the emerging world, demographic growth and urbanization were certainly critical factors supporting demand growth for electricity but they were not the only one. Energy-intensive industrializations, specific trajectories of economic growth and increased utilization of electrical equipment, particularly for air-conditioning, are other reasons which contributed to the steadiness of demand growth.

MENA countries share several common features: exercise of political power, implicit or explicit contracts between social structures and political leaderships, cultural and religious links. . .etc. These features directly or indirectly impacted the governance and the organisation of the strategic and ubiquitous electricity sector. However, these countries can differ widely in terms of economic and human development standards, sources of revenue, private sector development, credit ratings, to cite a few. At times, sub-regional or countries experiences may contradict regional trends. This diversity can be used to gain a better understanding of factors and dynamics at play. The presentation of the first three experiences in the region (Morocco, UAE and Oman) at the end of this chapter will be helpful in that regard.

The 2000–2008 period was the most fertile in terms of efforts for electric-sector reforms, and coincided with escalating oil prices, in addition to profound changes in the region, at the macroeconomic, social and political levels. All the electricity sector structures today in MENA were created during that period.

2.1 The Limits of Generic Modelling: High Demographic Growth, Rapid Urbanization and Industrialization

The pre-1990 population boom is having important implications. Simply put, since the end of the 1990s, a growing number of young adults are starting professional and marital lives, requiring new houses and new jobs, and creating a multiplier effect on energy demand growth. In his 2009 essay "Education vs. Extremism", Sheikh Mohammed bin Rashid al Maktoum, Vice President and Prime Minister of the United Arab Emirates, and Ruler of Dubai, identified unemployment, and more particularly unemployment of the youth, as a major problem affecting Arab countries.[1]

[1]"More than half of the 300 million residents of the Middle East are people under 25 years of age. The region has the fastest growing labor force in the world. With an already high unemployment rate of 15%, the Middle East must create 80 million new jobs in the next 5 years just to keep apace of our demographics. Unemployment is a problem afflicting all 22 member states of the Arab

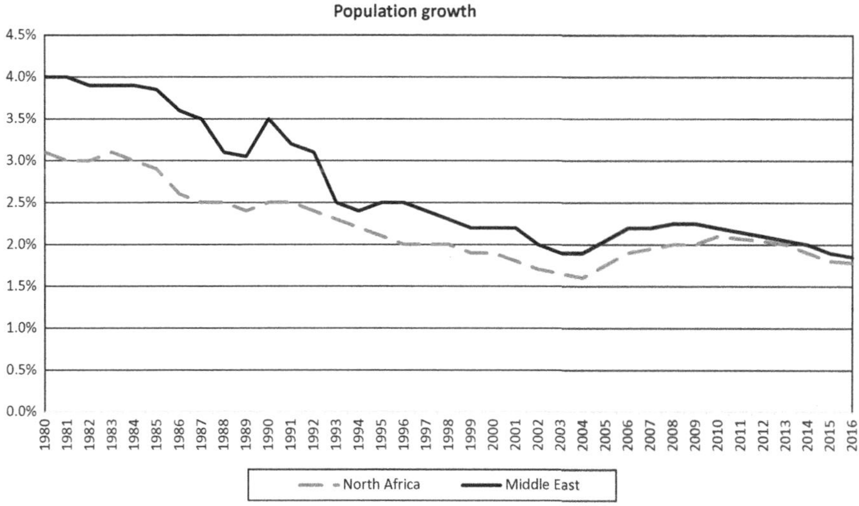

Fig. 2.1 Demographic growth rates in MENA 1980–2016. Source: World Bank, International Division of the US Census Bureau

Demographic growth has been slowing down, similar to the transition which occurred first in Western countries and then in Latin America and Asia. For example, the number of births in Iran dropped from more than 2 million a year in the mid-eighties to around 900,000 recently. Therefore, the effect of historic demographic growth should continue to impact electricity demand for a few years, but the acceleration is already slowing down (Fig. 2.1).

In fact, electricity consumption grew much faster than demographic growth. Gulf countries have often been singled out as registering some of the highest electricity consumption per capita in the world. The earlier benchmarking efforts, dividing total electricity consumption by total populations, failed to capture the underlying dynamics of the region. The following chart provides the number of kWh assumed consumed per capita. Gulf countries with disproportionately smaller populations registered the highest calculated electricity consumption per capita. Similar charts were often used in international conferences or meetings, notably in burgeoning climate change discussions (Fig. 2.2).

League, but it is most conspicuously a youth issue. Fifty percent of the jobless are under the age of 25, roughly double the world average. Women have an especially difficult time finding jobs. These increasingly restive youths are particularly vulnerable to those who would preach radicalism and hostility toward the West, especially the U.S. Why did Arabs fail to make deep structural reforms in education and in stimulating employment opportunities? The Arab world's track record on education, particularly girls' education, is discouraging. Sixty-five million adult Arabs are illiterate and two-thirds of them are women. More than 10 million Arab children between the ages of 6 and 15 are still not enrolled in any schooling, and on current trends this number will increase by 40% over the next decade. This is a monumental waste of human capacity." H. H. Sheikh Mohammed bin Rashid al Maktoum, *Education vs. Extremism*, the Wall Street Journal, June 3 2009.

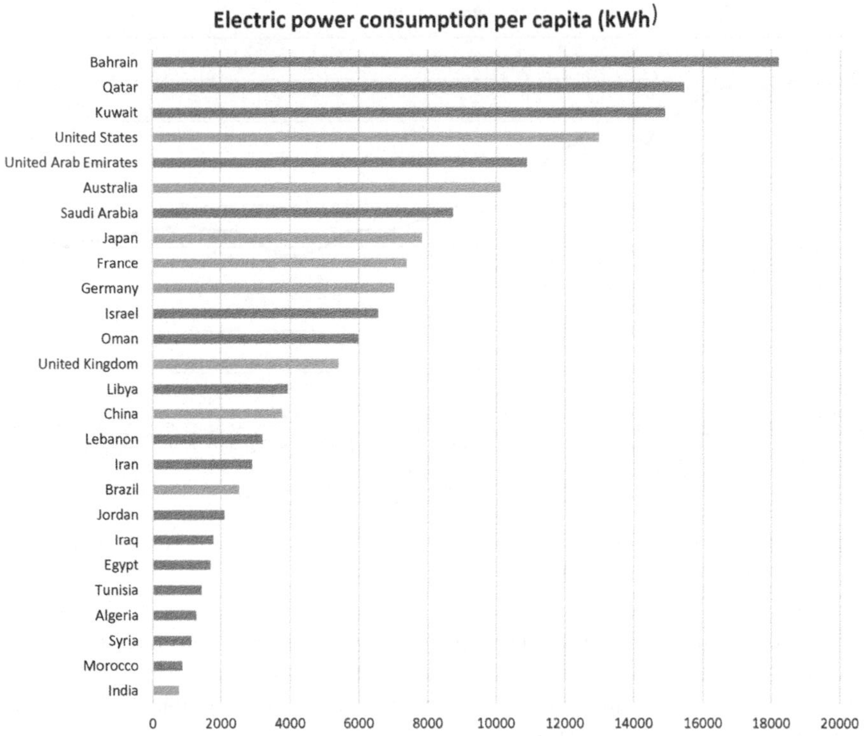

Fig. 2.2 Electricity consumption per capita in MENA and selected countries. Source: data from World Bank

In the mid-2000s, it became more common practice to compare *residential* electricity per capita, given the disproportionate electric consumption of other sectors, e.g. industries, in some countries.

In retrospect, this benchmarking glitch provided interesting insights and contributed to develop more research on energy demand in the region. Micro-economic elements (consumption behaviours, lifestyles, building codes, efficiency standards. . .) became central in understanding the regional dynamics. Subsidies, or the more politically correct worded 'incentives', were being scrutinised, but challenging them turned out to be a complex exercise.

As mentioned earlier, demographic pressures created social challenges, such as the urgent need to create jobs and employment on a large scale.[2] The high demographic growth of the pre-nineties created a generation of young adults who, today, would need employment, infrastructure, housing and various utilities. Governments started to explore alternative ways to monetize their energy resources, and sought to

[2]See for example Nabli & Keller « The macroeconomics of labor markets outcomes in MENA over the 1990s: how growth has failed to keep pace with a burgeoning labor market », World Bank, June 2002.

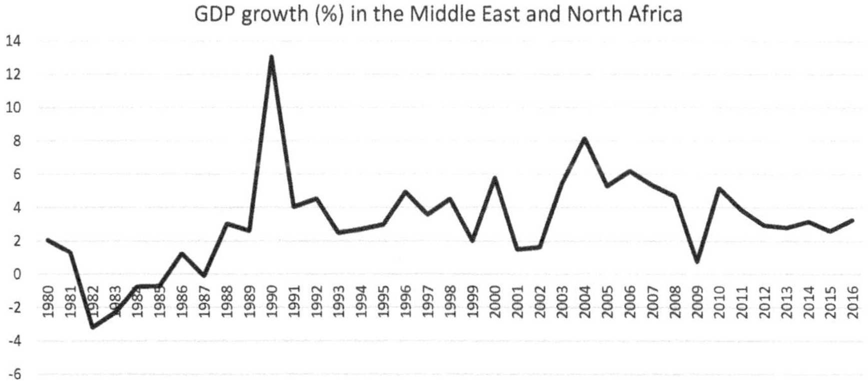

Fig. 2.3 Real GDP growth rates in MENA 1980–2016. Source: World Bank

create more jobs domestically. Some countries embarked in ambitious industrial plans, supported by below-market energy prices (fuel and electricity). In parallel, electricity residential tariffs have been widely subsidized in most, albeit not all, countries in the region (Fig. 2.3).

It is generally perceived that economic growth has a direct and proportional impact on energy demand. Energy modellers have been building regressions and econometric models linking economic and demographic growth to energy and electricity demand. This is another generic framework which fails when applied to most MENA countries.

Economic growth in the region is "distorted" by international oil price movements and then by the contribution of the hydrocarbons' sector growth to the economy. In modelling exercises and energy demand forecasts, it is more appropriate to consider the growth in the non-hydrocarbons' sector if data is accessible. Moreover, because of the subsidized energy prices in domestic markets, movements in international energy prices do not affect the energy demand of the country, while having a profound impact on the government's budget and ultimately on the country's economic output.

KAPSARC's research (2016) on the long run price elasticity of electricity demand for Bahrain, Kuwait, Oman and Saudi Arabia finds it to range from −0.16 to 0 suggesting that the electricity consumers' response to price changes is very limited, or that demand is very price inelastic. The income and population elasticities are also found to be inelastic—but generally greater than the price elasticities. The long run income and population elasticities of electricity demand for the four countries are found to range from 0.4 to 0.9 and 0 to 0.8 respectively.

In some oil producing countries, electricity consumption by the upstream sector has also been on the rise due to the increasing number of mature oil provinces and the need to use additional energy-consuming technologies to extract more oil. Energy intensity of upstream operations increases as oilfields deplete and as new reserves are discovered in more complex reservoirs. This last category of demand is however difficult to quantify and usually dealt with distributed generation in situ (generally

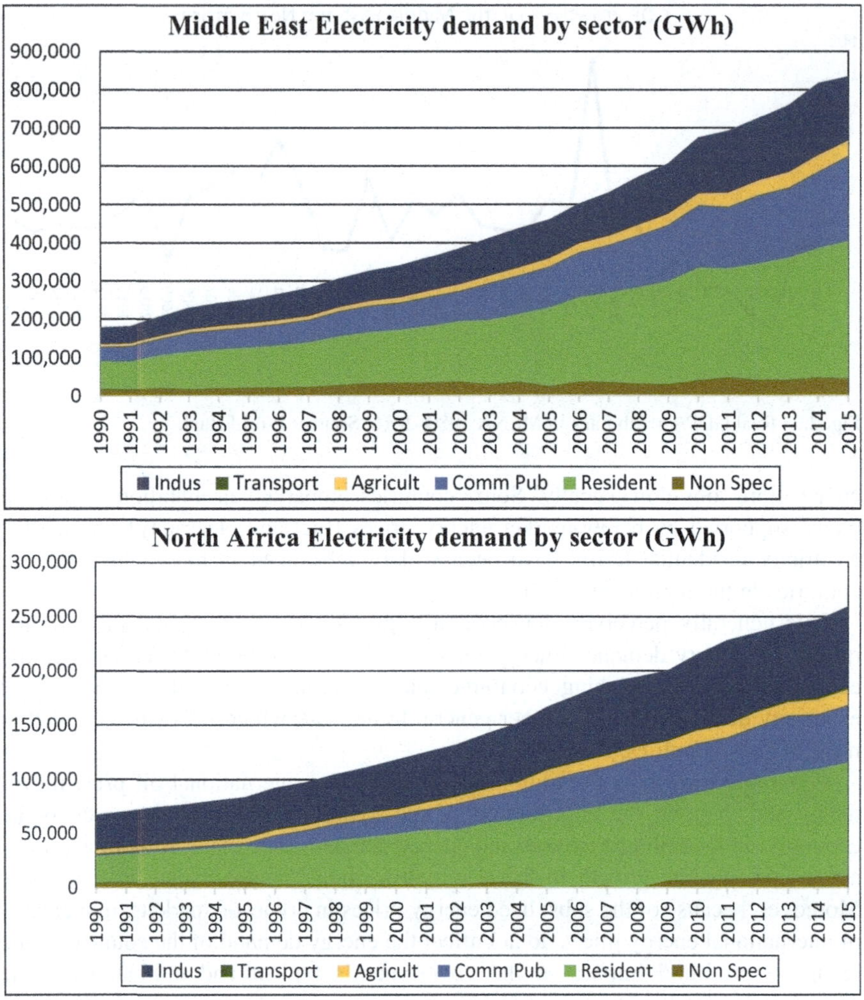

Fig. 2.4 Electricity demand by sector in MENA. Source: The International Energy Agency (IEA)

thermal—oil, diesel or gas-fired, with old oilfield power systems using low-efficient technologies).

The energy importers in the region, vulnerable to oil prices fluctuations, gradually adapted domestic energy tariffs—closer to market prices. Their domestic markets are less "insulated" from international energy prices fluctuations.

Figure 2.4 summarizes historical electricity demand by consuming sectors. Residential demand has been growing relatively fast. Today, with 6% of the world's population, MENA electricity demand accounts for less than 4% of global electricity consumption. Also, there is a wide discrepancy between North African and Gulf countries. Population sizes play a major role in generating these differences but other factors also come into play. Weather is a prime example, since in the Gulf countries,

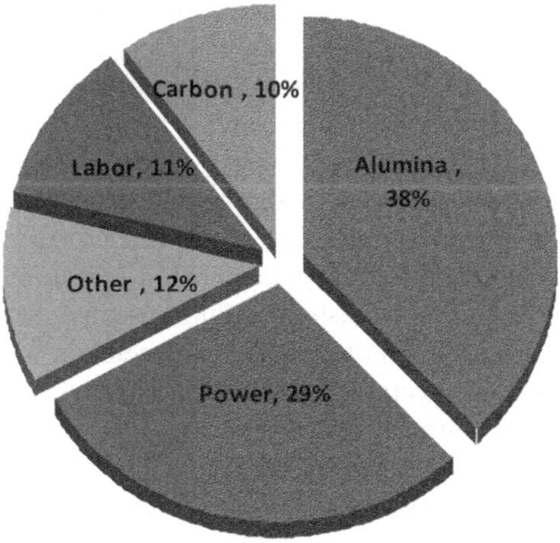

Fig. 2.5 Typical Aluminium input cost structure. Note: Based on a typical Western Europe smelter costs; labour share is very region-specific. Source: Commodities Research Unit (CRU)

air conditioning is the largest single element of residential electricity consumption (60% in Dubai, 70% in Saudi Arabia).

Demographic growth, also induced by increased numbers of expatriate workers, urged governments to seek job creation domestically.[3] This phenomenon coupled with the increased need to optimize the monetization of natural gas led these countries to launch large industrialization programs, either through direct government spending or through attractive terms for the private sector. Given the low gas prices in the region, the countries attracted large electricity intensive industries (aluminium—see Fig. 2.5), gas-fed industries (petrochemicals) and other industries using natural gas as a fuel in their process (cement). Over 9% of total global primary aluminium production is expected to come from the region by 2020.[4] Long-term cheap gas contracts were the primary reasons behind international investors' interests, encouraging these export-oriented projects.[5] For example, Alba's smelter secured a 20-year gas supply contract at $0.5 per MMBTU when it started operations in 1971. More recent entrants secured similar below-market gas prices of $0.75–1.30 per MMBTU. These smelters are among the lowest production cost bases in the world, counting on gas price, or the equivalent electricity cost through Energy Conversion Agreements (ECAs), as their most significant competitive advantage.

The sustainability of such a pace of industrialization, driven by economic diversification imperatives, has been heavily questioned. While MENA economies are already perceived as highly vulnerable to global markets' conditions because of

[3]Almost every country in the region, particularly Gulf countries, had a program of nationalization of the workforce. Foreign companies willing to invest in the country have to hire and train a quota of local citizens. See for example Kapiszewski, "Nationals and Expatriates: Population and labour dilemmas of the Gulf Cooperation Council States", Ithaca Press, 2001.

[4]"The Middle East as an emerging hub", Walid al Attar, Emirates Global Aluminum, Platts Aluminum Symposium, January 2014.

[5]85% of the output is exported outside of the region.

hydrocarbons exports' dominance, the focus on energy-intensive industrialization is also viewed as increasing that vulnerability.

In any case, the industrial sector contributed heavily to the growth in electricity demand in the region, at least on a baseload basis. The extreme temperatures witnessed in Middle Eastern summers, particularly in the Gulf countries, led to runaway demand, and exposed transmission bottlenecks. Therefore additional investment in transmission are needed (see Appendix A for a comparison of peak summer dry bulb temperatures in Middle Eastern cities vs. examples of US cities with successful district cooling systems). In the long term, these investments would lead to more efficient use of generation assets and security of the system during extreme operating conditions. However, several utilities are still struggling to finance these investments.

Finally, it is important to note the propensity of major hydrocarbon-exporting countries to invest revenues domestically or in the region, instead of "recycling petrodollars" in western economies. This trend accelerated after 2001–2002.[6] In absolute terms, Saudi Arabia invested the most in capital expenditures internally. In relative terms, Qatar has been the most aggressive relative to its hydrocarbons' revenue.

These investments created the foundation and the momentum for a steadier growth in energy demand in general, and electricity demand in particular. In various countries, large "industrial zones"[7] and "economic cities"[8] were created, in addition to the infrastructures associated with them. This will continue to support a minimum level of electricity demand growth during at least a large part of the current and forthcoming decade.

2.2 Future Investment Needs

There are several methods to forecast future electricity needs. Since electricity can still not be stored economically, load forecasts, the most accurate possible, are critical for all the market players, including the participants in electricity generation, transmission and distribution. There are broadly three different categories of

[6]The "conventional wisdom" attributes this change to 9/11, but there are no consistent data to prove that Middle Eastern countries pulled out a large part of their money of the western banking system after 2001 because of that.

[7]Large industrial zones include Ras Laffan and Messaied in Qatar, Jubail and Yanbu in Saudi Arabia, Ain Sebbaa and Jorf Lasfar in Morocco, Sohar in Oman.

[8]Saudi Arabia is developing five economic cities by 2030 ($80 billion of estimated total investments), hubs for development and growth. Among the five economic cities by 2030, Jizan is probably the most spectacular: an industry-focused city with $30 billion of Saudi investments. Foreign investors can take part in several projects within the city. Moreover, the King Abdullah Economic City (KAEC), on the Red Sea coast near Rabigh, is the first project to have been launched (in December 2005).

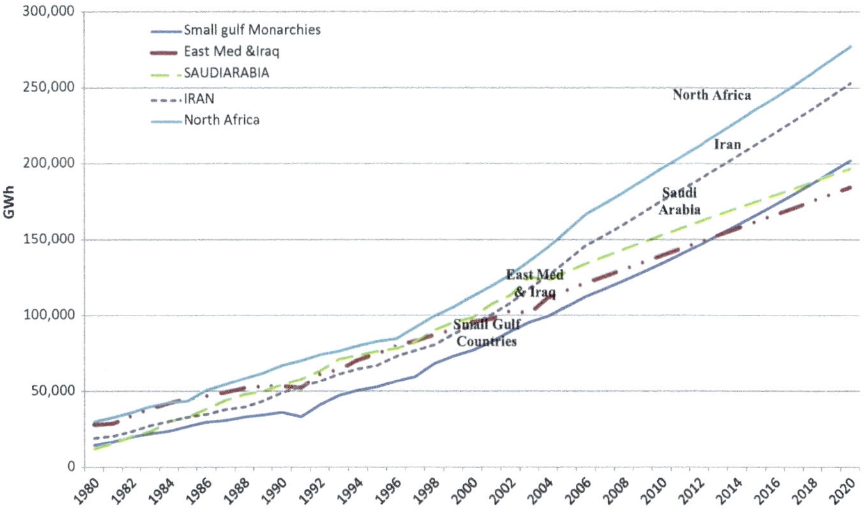

Fig. 2.6 Electricity demand growth in the region to 2020

forecasts: short-term (1 h–1 week), medium (1 week to a year), and long-term. These would use three broad categories of forecasting techniques: traditional techniques, modified traditional techniques and soft computing techniques.

Long-term forecasting can provide a projection of future capacity and investment needs, using "top-down" macroeconomic indicators—e.g. GDP and population—as input variables, and/or "bottom-up" building up on detailed analysis of every sector of demand. Extrapolating industrial demand for instance, based on the industrial plans of the country and an assessment of the industrial projects most likely to be commissioned, requires a detailed understanding of demand patterns of the various industries and of the macro-economic environment of the country.

An econometric equation with endogenous variables (independent) such as population and GDP and the exogenous (dependant) variable as electricity demand enables to forecast electricity demand. If consistent data could be accessed, non-oil GDP as an independent variable instead of overall GDP is preferred, in order to avoid the distortion created by large oil price fluctuations for oil-exporting countries:

$$\mathrm{Di} = \alpha^{*}\ \mathrm{POPi} + \beta^{*}\ \mathrm{GDPi} + \gamma;$$

Di: Electric demand for year i
POPi: Population for year i
GDPi: Gross Domestic Product for year i
α, β and γ being the parameters of the equation.

There appears to be a strong enough correlation between the different variables and electricity demand. A multiple linear regression to forecast an approximate value of future electricity demand, based on third-party data and scenarios for population and GDP growths, using the method of the least squares, provides the following forecast (Fig. 2.6).

It is important to note the large seasonal—sometimes intraday—demand swing witnessed in the region given the extreme temperatures and the continued acceleration of demand experienced during the first part of the projection period (at least up until the global recession of 2009).

It is then possible to estimate the amount of total capacity needed in the region; five to seven years being a reasonable planning period for power projects. To translate these projections into capacity needs, other assumptions will be needed, e.g. 15–25% of system's losses and 5–15% reserve margins depending on the country, in addition to the amount of generation capacity that will need to be retired during the period. In reality, the international energy outlook (2016) of the EIA estimates that the actual installed capacity in the Middle East in 2015 was at 258 GW. A large share of the difference is explained by war-affected countries such as Iraq where installed electric capacity could not exceed 10 GW (Table 2.1).

Table 2.1 Electricity capacity in MENA

Country	Installed capacity (GW) 2006	Peak demand (GW) 2006	Peak demand Growth (%)	Capacity needed in 2015 (GW)	As % of 2006 capacity	
					Oil	Gas
Middle East	175.0			267.2		
Bahrain	1.4	1.3	7	2.5	13	87
Iran	40	35.5	6	67.5	28	59
Iraq	5.4	8.9	NA	15.5	57	14
Israel	10.9	9.4	2	16.4	6	18
Jordan	1.9	1.9	10	3.3	21	78
Kuwait	9.5	8.7	7	16.8	55	45
Lebanon	2.3	2.3	4	4.0	90	0
Oman	3	3	5	5.8	33	67
Qatar	3.4	3.2	7	6.2	0	100
Saudi Arabia	32	27.3	5	46.6	59	41
Syria	7.5	7	4	12.2	21	62
UAE	18	10.6	8	20.4	9	91
Yemen	0.9	1	4	1.9	100	0
Turkey	38.8	27.5	6	48.0	4	44
North Africa	39.4			64.8		
Algeria	9.27	7	5	12.9	7	87
Egypt	17.7	17	4	31.4	27	55
Libya	4.6	4.8	10	8.9	40	60
Morocco	5	3.9	6	7.2	0	40
Tunisia	2.8	2.4	7	4.4	34	66

Source: Author's calculation based on data from local Sources, MEES, EIA

Table 2.2 Electricity capacity, demand growth rates and fuel mix in MENA 2006–2016

	Installed capacity (GW) 2016
Bahrain	3.9
Iraq	32.1
Jordan	4.6
Kuwait	18.8
Lebanon	3.0
Oman	7.8
Qatar	8.5
Saudi Arabia	74.7
Syria	9.6
UAE	28.7
Yemen	1.5
Algeria	19.0
Egypt	39.1
Libya	10.2
Morocco	8.2
Tunisia	5.5

Source: Arab Union of Electricity

Therefore, even in high-growth markets like MENA, system planners can build with a certain level of confidence capacity needs forecasts over a reasonable planning timeframe, barring unforeseen events (Table 2.2).

On the other hand, capital costs great variations (inflation) over the 2000–2008 decade made it difficult to give a precise figure for actual future investments.[9] There are several drivers behind these variations. Demand changes for turbines and boilers affect capital costs. Steel costs increases have a direct impact, as mills pass through higher raw material costs particularly for iron ore, coking coal and fuel. Using current projects' costs as benchmarks, one can assume the following approximate overnight capital costs[10]: $800 per kW for an oil-fired plant, and $1000 per kW for a combined-cycle gas turbine. By way of comparison, nuclear would cost 4600 $ per kW, wind 2000 $ per kW and supercritical coal 2700 $ per kW.

Therefore, the cost of installing the 100 GW needed in the whole region—between 2006 and 2015—would exceed $91 billion, that is to say around $10 billion per annum.[11] This investment figure does not include new transmission or distribution lines stemming from the connections of new villages (rural electrification) or new cities (newly built), or from grid reinforcement. The IEA estimates that for each

[9]CERA's capital cost index shows that power capital costs nearly doubled between 2000 (100) and 2007 (194). It even exceeded 230 during the summer of 2007. It reaches 171 in 2007 if nuclear is excluded.

[10]Based on empirical evidence from projects' costs and on CERA's Capital Costs Forum data.

[11]By way of comparison, the Arab Petroleum Investments Corporation (APICORP) estimates MENA's capacity expansion needs in the MENA region at 60 GW between 2007 and 2011, at a cost of $61 billion. MEED estimates that 113 GW are needed between 2007 and 2015, at a cost of $109 billion.

Table 2.3 MENA OPEC oil export revenue

	2002	2004	2006	2007	2008	2010	2012	2014	2016
MENA OPEC	157.8	280.9	485.6	538.7	763.3	606.1	987.3	740.9	357.2
WTI $/barrel	26.18	41.51	66.05	72.34	99.67	79.48	94.05	93.17	43.29

Source: OPEC. In billion USD

kilowatt of new capacity added in MENA, accompanying investment in networks (transmission and distribution) will be almost $850 per kW. Therefore, the installation of 100 GW would roughly need an additional $85 billion in investments in transmission and distribution, or an additional $9 billion per annum.

By way of comparison, the total annual investment needed is between 2% and 5% of the region's oil export revenue during the period of high oil prices (2006–2016) (Table 2.3).

Demand-side and supply-side techniques could help decrease the investments needed, or at least delay investment decisions where possible, e.g.: energy efficiency and conservation programs, dispatch optimization of existing power generation, wholesale power purchase or trading, time-of-use pricing (on-peak usage reduction), smart metering (giving end users access to real-time usage data, to reduce consumption especially when combined with time-of-use pricing), smart grid technologies.

The issue is most of these techniques assumes the existence of a wholesale market with the right regulatory framework. The last three strategies require a strong regulatory framework, an advanced and sophisticated market and additional investments to be implemented.[12]

It is important for all countries to implement efficiency programs and techniques. These programs will come at a cost—no negative costs as such—.[13] However, they could contribute to the development of sound regulatory frameworks, and more importantly, they will send signals to the customer on the value of energy.

[12]In the United States, the Obama administration announced in April 2009 the distribution of $4 billion in federal grants to start upgrading the electrical grid and introduce smart grids. This is part of the $787 billion stimulus package passed in February 2009.

[13]In reference to the 2007 McKinsey study, *Reducing US greenhouse gas emissions: how much at what cost*? This study concluded that GHG reductions opportunities are widely scattered across the US economy, and that 40% of abatement options can be achieved at negative costs (the concept of negative costs is debatable). But more importantly, reducing the carbon intensity of the power sector and improving the energy efficiency of buildings and appliances are the two most effective approaches out of the five clusters identified in terms of Gigatons of CO_2 avoided. The third cluster is pursuing a range of targeted measures in energy intensive sectors of the industry. The last two clusters are higher energy efficiency and lower emissions in transportation (vehicles) and expanding/enhancing carbon sinks.

2.3 The Three First Experiences, the First Lessons: Morocco, UAE and Oman

MENA can also be divided into two very distinct types of countries and markets: hydrocarbons resource-poor countries (usually heavily populated) and resource-rich ones. Among resource-rich countries, some were able to build a non-oil economy with a significant industrial base (Egypt, Iran). In the early nineties, the model in place in all these countries was that of a vertically-integrated monopoly with some industrial auto-producers (Office Chérifienne des Phosphates in Morocco, Dubal in Dubai, Alba in Bahrain, SWCC in Saudi Arabia) generating their own needs and injecting the power surplus into the grid.[14]

Who Started?

Oman (al Manah, the region's first BOT project awarded in 1994), Morocco (Order in Council ending ONE's monopoly on generation promulgated in 1994 and Jorf Lasfar commissioned in 1995) and Abu Dhabi (regulatory framework established in 1997 and privatization started in 1999) were the vanguards. Having started to introduce BOT schemes, these countries benefited from the experiences in OECD countries and from the nascent ones in Latin America or Central Europe. The three countries were the first ones to introduce legislations to reform their electricity sectors. The Moroccan example slightly differs because this country is a hydrocarbons importer and this peculiarity has strongly influenced (a) the specificities of its assets and (b) the choices of governance structures (Williamson). Morocco has tried to move beyond the simple Single Buyer Model by introducing coexisting regulated and deregulated markets. The project was approved by the government, but the draft law which foresees the opening of the sector one year after ratification, has stalled in Parliament.

The Moroccan Experience

Before the independence, the electricity sector was managed by a vertically integrated monopoly, the French Energie Electrique du Maroc (90% of national generation). Distribution was handled by private firms. After the independence, electricity was defined as a strategic sector. EEM's concession was ended and transferred to the Office National de l'Electricité (ONE), created in 1963 as a monopoly for generation and transmission. ONE has been administratively and technically supervised by the Ministry of Energy and Mines.

[14]Dubal and Alba are aluminum companies while SWCC (Saline Water Conversion Company) is a state-owned desalination utility in Saudi Arabia.

The Moroccan electricity sector experienced series of crisis after the eighties, because of the combined effect of drought, increased call on thermal generation, demand growth and delays in financing and implementing projects. Technical issues related to the reliance over relatively cheap and flexible but weather-related and highly unpredictable hydro-power for an important part of supply. In parallel, production costs were closely linked to international fuel prices, and therefore volatile, while tariffs were not designed to take into account peak and off-peak or seasonal conditions. The increasing difficulty of ONE to finance or secure financing for the needed projects created an urgent need to find alternative solutions.

The Moroccan electricity sector suffered from the pitfalls of a monopolistic structure (delays, bureaucracy, limited finances) without fully benefitting from the introduction of the private sector in distribution (because of lack of coordination and no direct regulation). Finally, the limited electricity coverage of the country[15] somewhat hampered the expansion of the sector, although this was subsequently addressed by the Programme d'Electrification Rurale Global (PERG).[16]

Market reform in Morocco started with the introduction of the private sector in the generation (1994) and in the distribution (1997). Moreover, independent generators (mines, phosphates...) could sell surplus electricity to ONE through negotiated conventions as they were connected to the grid. Distribution was handled by 12 "régies" placed under the guardianship (tutelle) of the Ministry of Interior, or under delegated management (gestion déléguée) by private operators in a few cities. Electricity tariffs were defined in the first instance by a Prime Minister's decree and secondly through contractual negotiations between the Commune and private operators. It was actually the Ministry of Interior who incited local collectivities to call upon private operators to manage local public services.[17] The main outcome from this situation was the end of the uniform national electricity tariff and the starting of differentiated tariffs (discrimination) between consumers. Morocco chose not to create an independent regulator.

Ironically, it is ONE who was asked to suggest solutions with legal amendments and to lead additional reforms. The company's Board Decision of 2001 paved the way for the current but long-delayed reform by partially opening the sector to competition and public sector modernization. Until the implementation of that reform, ONE (l'Office National de l'Electricité) would have been a vertically-integrated monopoly responsible of a minimal part of generation, distribution all

[15]Bouayad (2001) compares 25% electrification in Morocco in 1993 to 86% in Algeria or 70% in Tunisia during the same year.

[16]« PERG allows to accelerate rural elerification and to commit ONE until 2006 to the mission of assuring the universality of the electric public service » Benhima (1999).

[17]The Urban Community of Casablanca was the first one to delegate the management and the distribution of electricity and water. It was followed by Rabat (Rabat-Témara-Salé).

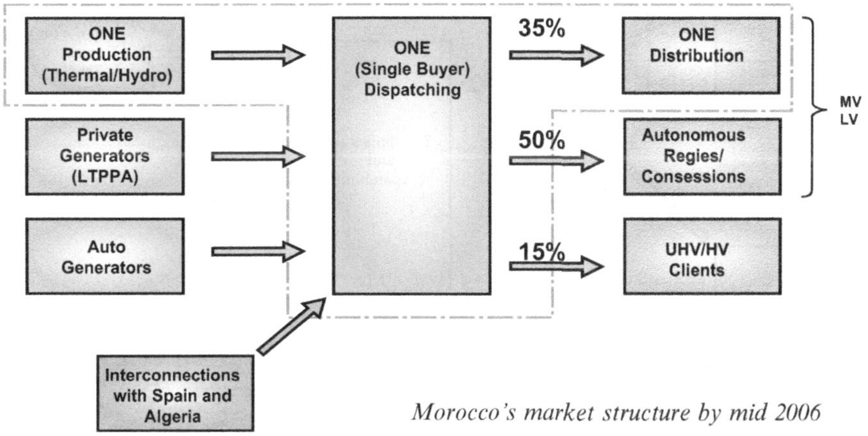

Fig. 2.7 Morocco's market structure by mid 2006

over the country except in five cities.[18] ONE is also the single buyer of the electricity generated by its own plants, by the Independent Power Producers[19] and of the electricity imported (mainly from Spain) or the surplus which is injected into the grid by auto-producers (Fig. 2.7).

The reform described in the ONE's Board Decision of 2001 aimed at the creation of coexisting regulated and deregulated markets was not implemented. A regulator, "Market Operator", would have been created as well as a "National Controller" to oversee the system. The principle of Third Party Access (TPA) to the network would be reinstated but the details of its functioning were not defined. Consumers' eligibility's threshold—based on annual consumption—would be fixed by regulation and decrease over time (during the first year, only Ultra High Voltage-High Voltage clients are eligible). Eligible clients would choose to stay in regulated market or not, under a set of conditions.

In the deregulated market, clients might purchase electricity through bilateral contracts, for short periods of ½ to 2 years or from a yet-to-be created spot market. Exchanges will be permitted between the two markets, but the Single Buyer's sales will be particularly monitored, in order to avoid monopoly/market power. In the regulated market, prices will still be defined by Prime Minister's decree and the single buyer still managed by ONE (Fig. 2.8).

---

[18]Three private autonomous régies are responsible of distribution in five cities: LYDEC (Suez 30%, ELYO 21%, Fipar Holding 21%, Wataniya 15% and stock exchange 14%) in Casablanca and Mohammadia, REDAL (Compagnie Générale des Eaux 51%, Compagnie Marocaine des Services à l'Environnement 49%) in Rabat, AMENDIS (Veolia Environment, HydroQuebec, ONA, SOMED) in Tangiers and Tetouan. Other cities in the country are serviced by state-owned régies (ONE) like RADEMA for Marrakech and RADEEF for Fes.

[19]Jorf Lasfar was responsible of 60% of the electricity generated in the country in 2004. This share decreased in 2005 with the commissioning of the Tahaddart power plant but Jorf Lasfar will remain a predominant player.

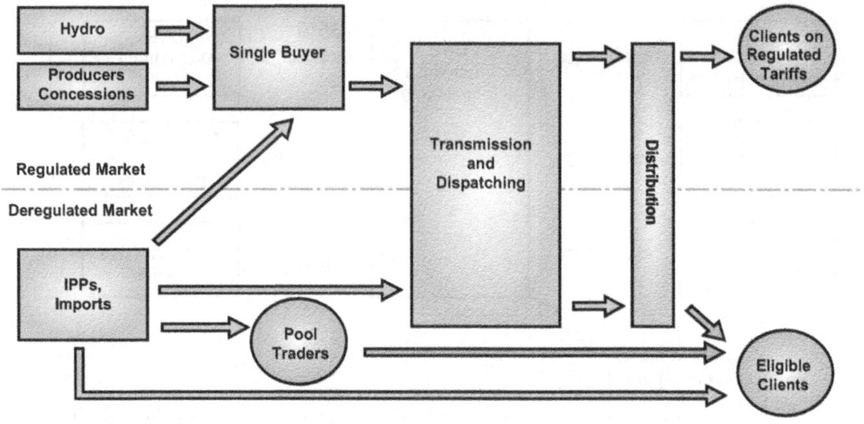

Fig. 2.8 Morocco's post-reform market structure (not implemented)

The regulated market would comprise producers with Power Purchasing Agreements—including Jorf Lasfar and Tahaddart-, peak units, wind farms and hydro units (protected from free market conditions). ONE (or its subsidiary) would retain monopoly over supply to non-eligible clients (in regulated market). A draft Law had yet to be presented after several delays to the Government's Council and to the Parliament, but the project was strongly objected by various stakeholders, apparently including the national utility.

In Morocco, tariffs have reportedly decreased by 30% since the mid-nineties, mainly to the benefit of industrial customers. Retail tariffs were left largely unchanged. Recent increases targeted mainly industrial customers and peak consumption.

However, the competitiveness of plants in terms of production costs in the deregulated market has been unclear, so was the probable impact (if any) of the possible additional reform on prices. A major decrease remains unlikely and the tariffs in regulated market might be lower than in deregulated market. Producers in the regulated market will not be able to access deregulated market in the latest version of the reform or under unclear conditions. All this still needed to be clarified in the final version of the Law, if the authorities decide to revive it.

The Emirati Example

The UAE is a federation of seven emirates. The power sector is supervised by government-owned authorities: the Abu Dhabi Water and Electricity Authority (ADWEA), the Dubai Water and Electricity Authority (DEWA), the Sharjah Water and Electricity Authority (SEWA) and the Federal Water and Electricity Authority (FEWA) for the northern emirates, the remaining four emirates. FEWA, DEWA and SEWA are vertically integrated and State-owned (Table 2.4).

Table 2.4 Size of the various electricity markets in the UAE

Installed capacities (MW)	Abu Dhabi	Dubai	Sharjah	Northern Emirates
2003	6186	5333	1698	1775
2015	15546	9656	2840	1703

Source: KAPSARC

The reform away from a monopoly model occurred first in Abu Dhabi.[20] The authorities decided to create a "Privatization Committee" for the power and water sectors, headed by Sheikh Dhiyab Bin Zayed al Nahyan, who will later become Chairman of ADWEA.[21] The reform officially started in 1997 and in 1998 with a new Law for the sector. It established various entities which were initially owned by the Abu Dhabi government, through the Abu Dhabi Water and Electricity Authority (ADWEA). The Emirate followed the Single Buyer Model and created the Abu Dhabi Water and Electricity Company (ADWEC) in charge of buying all the electricity and water produced. A limit on foreign ownership was set at 40% on all Brownfield and Greenfield IWPPs. As of end 2006, there were eight generation companies in Abu Dhabi, two of which were still State-owned (and candidates for privatization). The transmission (Transco) and distribution (2 companies in Abu Dhabi and Al Ain) parts were unbundled but remained State-owned. The sector is supervised by the Abu Dhabi Water and Electricity Authority (ADWEA) chaired by Sheikh Diab bin Zayed al Nahyan, and regulated by the Regulation and Supervision Bureau (RSB). Dubai, on the contrary, has been experiencing a strong internal resistance to market reform (Fig. 2.9).

In 2009, four IWPPs were launched with a total contracted output of 5000 MW and 255 MMgd, and about $5 billion of commercial financing. In 2003, ADWEA decided to partner in the long-standing Sohar aluminium smelter project (with 1100 MW captive power plant) in neighbouring Oman. But ADWEA's remit was not to go overseas. The government created later the Abu Dhabi National Energy Company (TAQA) for that purpose (see Sect. 2.3 for more details on TAQA). At the time, Sheikh Diab's stance was that ADWEA's "going overseas will be assessed on a case by case basis, but at the moment we don't see ourselves going anywhere else". The next IWPP, involving the acquisition and expansion of the Umm al Nar plant (Arabian Power Company), was launched in difficult market conditions: in the aftermath of 9/11, the Enron scandal and the bursting of the gas bubble in California. ADWEA received only two bids, but with surprisingly good offers.

Plans to sell part of the government's 60% stake through an IPO to local investors has always been in the agenda, but so far these plans have not materialized. Moreover, the plan to privatize the two distribution companies has been facing two main challenges. First, because reserve margins for power and water had been

[20]See Benali (2002).

[21]Sheikh Dhiyab bin Zayed al Nahyan is the 17th son of the late Sheikh Zayed, former President of the UAE and ruler of Abu Dhabi. Shaikh Dhiyab has also been the Director of the UAE presidential Diwan (consultative council).

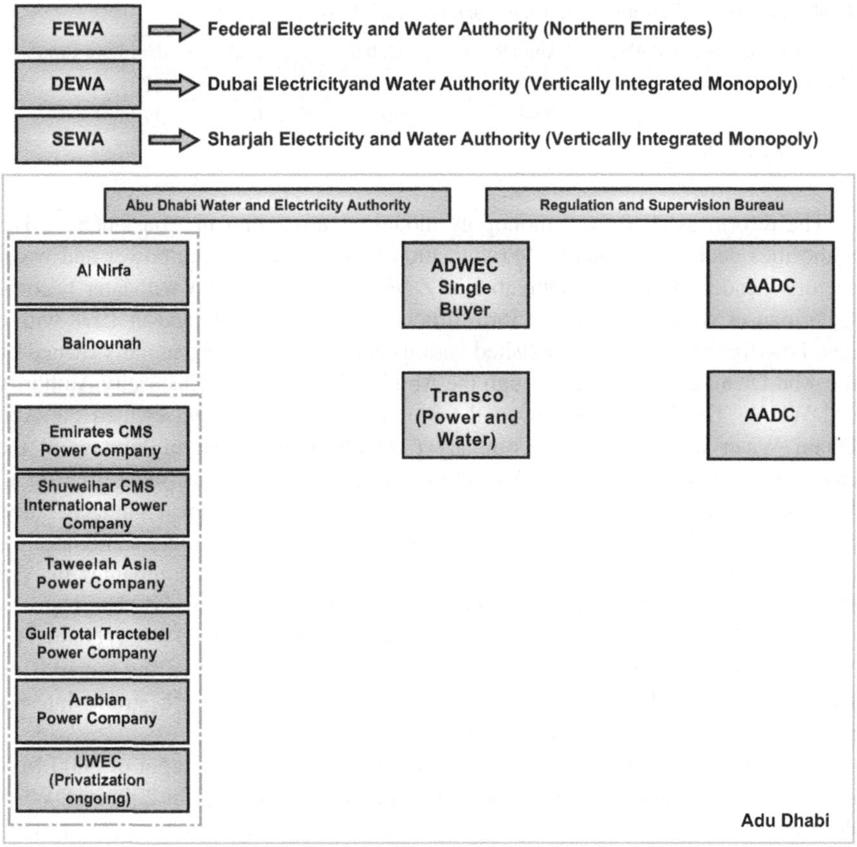

Fig. 2.9 The Electricity sector structure in the UAE

extremely tight during the first years of the reform program, generation had been given priority. Second, during the early days of the reform, there were no national power and water networks. Since then, reserve margins improved because of the series of IWPPs launched and more than $2 billion were invested in power and water networks ($545 million in the electricity grid and $1 billion in the water network between 1998 and 2003). But financial investments are not sufficient to reform the companies to privatize. In 2005, Abu Dhabi Distribution Company (ADDC), also chaired by Sheik Dhiyab, embarked in a 5-year improvement program in order to improve performance of the company, following review by external consultants.[22] During the same year (the last publicly available annual report), ADDC's levels of

[22]The first step was to start a process of internal restructuring to separate activities into "asset owner/ manager" and "service provider", to focus greater attention on asset management and investment planning while driving down operating costs and improving levels of service.

debt (AED 1.7 billion, $0.4 billion) were still high and actually increased, due to a deteriorating revenue collection performance from government departments. ADDC advocates the change of its disconnection policy.

In parallel, ADDC and the Regulation and Supervision Bureau (RSB) set up a new regulating control for the 2006–2009 period, using the RPI-X method, while tariffs are set by the Government of Abu Dhabi in conjunction with ADWEA and the RSB.[23]

The Omani Experience

In Oman, a new Law was enacted in August 2004 to implement policies approved by the Council of Ministers in 1999. Actually, the first IPP project in Oman started generating electricity in 1996. Prior to August 2004, the Ministry of Electricity and Water had been responsible for the whole sector. The sector evolved from "a self-regulated vertically integrated ministerial utility to an independently regulated unbundled commercially-based sector".[24] The responsibilities of the MHEW were transferred to the newly formed successor companies: the Electricity Holding Company, Oman PWPC, Oman Electricity Transmission Company, the three distribution companies, the Rural Areas Electricity Company and the generation companies. Oman Power & Water Procurement Company was established as a closed joint stock company (SAOC) in 2003 (Fig. 2.10).

The demand growth of 7–8% in recent years has been supported by the government macro-economic policy (major industrial developments in Sohar and Salalah, major tourism related projects and city planning initiatives in Duqum for example). But expected demand growth was not the only reason behind Oman's active reform and liberalization.

The State contracted in 1999 a group of advisors led by ABN AMRO bank, with Denton Wilde Sapte providing legal services and Mott MacDonald (UK) acting as technical consultant. Privatization of the State assets has been supervised by a Privatisation Committee which consists of eight ministers led by Sultan Qaboos. According to the advisory group in early 2002, "privatization of existing assets and getting new power and water desalination plants to be built privately was to allow the State to save RO 750 million ($1923 million) from its capital budget over ten years. In addition, increased efficiency through private involvement would lower costs and enable the state to cut subsidies without affecting tariffs set for end-users". Then, the sector reform was driven behind the scenes by the powerful Ministry of National Economy, probably to make sure that the country's economic development plans

---

[23]All controls, with separate price controls for Water and Electricity, are meant to last 4 years. All Income from licensed activities is to be included within the definition of "regulated revenue". Procurement charges and transmission service charges for both water and electricity are passed-through.

[24]John Cunneen, Executive Director, Authority for Electricity Regulation, Oman, POWER-GEN Middle East 2006—Abu Dhabi, 30 January 2006.

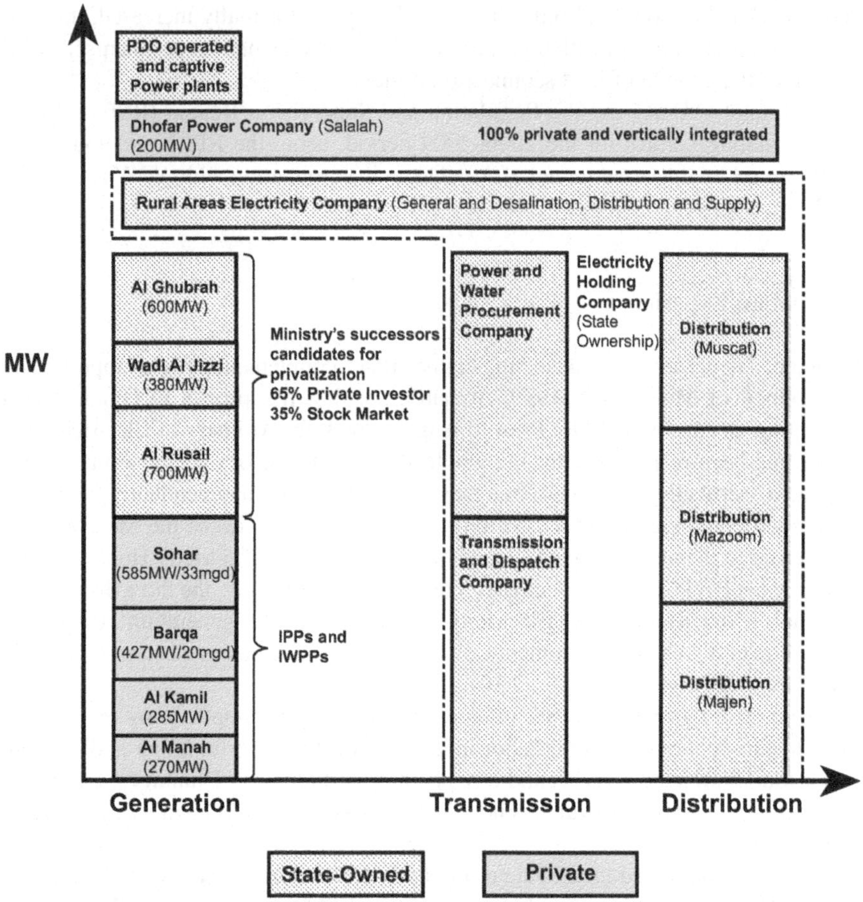

Fig. 2.10 The Electricity sector structure in Oman as of 2007

would not be hampered by a limited electricity sector. This drive even influenced the appointment of some key personnel. For example, Mr. Mohammed bin Abdullah al Mahrouqi was seconded by that same ministry and appointed as acting CEO of Power and Water Purchasing Company (PWPC). In 2007, he was named by Royal Decree Chairman of the new Public Authority for Electricity and Water with the grade of an Undersecretary.

Licenses to operate in the electricity sector were granted in May 2005. As of 2007, the electricity sector in Oman consisted of the Main Interconnected System (with seven generators—of which two have a generation and desalination licence), the Rural Areas Electricity Company and the Dhofar Power Company (a vertically integrated private company to serve the southern part of the country). The single buyer is the Power and Water Purchasing Company (PWPC) is expected to remain State-owned. A regulator (the Authority for Electricity Regulation) was created in

May 2005 and there are three distribution companies that are candidates for a future privatization. Apparently, and contrary to regional trends, the government has been thinking about privatizing the transmission company as well.

To date, on the generation side, all ventures are 100% privately owned as the sale of Rusail marked the first step of the 2nd phase of the privatization program. Oman was able to attract strong international power companies to the country (AES, Marubeni and Suez). The first four IPPs had the obligation to offer at least 35% of shares on the Muscat Securities Market within 4 years of incorporation. The Rural Areas Elec. Company, PWPC and Electricity Holding Company (an umbrella company) will remain State-owned.

The way AER regulates the sector is slightly different from the regulation in Abu Dhabi, although the general principles are similar. AER issues licenses or exemptions for market activity. It monitors performance and calculates the Bulk Supply Tariff. It sets the revenue cap for the transmission company (Maximum Allowed Revenue) and uses a combination of price cap and revenue cap for distribution companies. The latter pay a cost reflective BST to PWPC, while PWPC buys electricity and water from the generators following the terms of the PWPAs. Of course, AER comes up with recommendations particularly regarding new tariffs but the Ministries of Finance, Economy and Electricity (actually the Council of Ministers) have the final say. For example, AER has been recommending the implementation of cost reflective tariffs for industrial customers that will take into account the Bulk Supply Tariff, plus transmission, distribution and administrative costs. The Authority recommended the implementation of this new tariff in January 2007 but the change needs to be approved by the Council of Ministers first.

The pricing regime has been transparent since the onset of the reform. For example, in 2006, the wholesale BST was set at 1.9 cents/kWh during the whole year except for the period from May to August (2.6 cents/kWh 10 PM–2 AM; 20 cents/kWh weekday 1 PM–5 PM; 7.8 cents/kWh Thursdays 1 PM–5 PM; 5.2 cents/kWh Fridays 1 PM–5 PM. Then, during the same year, the Single Buyer estimated cost of electricity purchase was 3.3 cents/kWh (no subsidy at the wholesale level). The economic cost of electricity supplied was estimated at 7 cents/kWh and therefore the subsidy to the electricity sector at 3 cents/kWh.[25]

It is only in September 2016 that the Council of Ministers could approve the application of Cost Reflective Tariffs (CRT)—i.e. no subsidy element—for large Government, Commercial and Industrial customers whose annual consumption exceeds 150,000 kWh. The new tariff came into effect on 1 January 2017.

Political consensus is critical for prices and subsidies to be revised. Consensus is actually a key feature of politics in Oman and in other Gulf countries. There is a reasonable amount of literature that analyzes the utilisation of consensus in the exercise of authority in Gulf countries. Eickelman (1985) for example analyzes the typology of monarchic rule in Oman and the utilization of religious credentials and

[25]This is just the direct subsidy and does not take into account the indirect subsidy of below-market gas price offered to generation companies.

of a culture of consensus to legitimize the power of the ruler. Crystal (1995) lists consensus, violence, secession and matrimonial alliances as the most important political tactics used in the Arabian Peninsula local politics in the eighteenth century. At the time, Kuwait was the only political unit. As various states were formed, consensus remained an important feature of Gulf countries' internal and regional politics. This strong tradition of consultation and consensus can slow major and highly visible reforms.

Over the last decade, the Omani electricity sector has undergone significant institutional changes. It is still inspiring decision makers in neighbouring countries and it is considered by many to be an example of the way in which private power and commercial incentives can be used to meet growing electricity needs. The electricity needs of the country continue to be met in a generally efficient and economic manner. The sector has been significantly restructured to provide more transparency and clearly focus companies on their strengths. Incentive mechanisms have been put in place; a regulator and an electricity authority have been established. The sector faces now the challenges of a maturing industry, but not yet fully integrated. Three electricity grids exist in Oman. The two largest grids (the Main Interconnected System—MIS—and Petroleum Development Oman—PDO—for upstream operations) account for the vast majority of the load. The MIS and PDO grids have a limited interconnection, the purpose of which is to provide backup supply if needed. The MIS serves commercial and residential customers in the north of the country, as many large industrial customers using their own captive power plants. There has been some technical coordination between the parties, but there is no overall planning/coordination body to ensure that overall electricity operations, supplies and demands of the electricity users of the respective grids are optimized. An overall integrating body would open various opportunities. These include more efficient dispatch of power plants, lower fuel consumption, lower levels of required spinning reserve, baseload capacity sharing and more limited seasonality swings. However, integration of these grids still requires aligning the different operating standards applied in each.

These examples and others in the region show the limits of price regulation when applied in an environment of heavily subsidized tariffs (Oman, the UAE).[26] Conversely, Morocco highlights the advantages and the limits of a weak or absent regulation in the framework of a liberalization process. In all these cases, the effect of the reform on efficiency and cost—decrease or increase—is barely felt by the final consumer: the transfer mechanism does not exist anymore or does not occur directly (see Chap. 3). The production cost fluctuation has only an impact on the government budget; the "triangle of opportunism"—government, companies, consumers—turns into a simple dual relation between a government and (a few) companies.

[26]Most countries in the region adopted the RPI-X regulation. That was notably the case in Abu Dhabi and Oman.

2.4 Taking Stock on the Status of Reforms

The MENA region is unique, unique by its hydrocarbons' resources endowment, by its demographic trends, by the rapid urbanization and industrialization it has witnessed and by its spectacular economic growth. Resource-rich countries explored ways to monetize their energy resources and sought to create more jobs domestically. Some countries embarked in ambitious industrial plans which were supported by below-market energy prices. In parallel, electricity residential tariffs have been widely subsidized in most countries in the region.

Therefore, it should not be surprising to find in this region the world's highest electricity demand growth rates and the highest consumption per capita. Obviously, there is a wide discrepancy between highly populated North African countries, hydrocarbon-rich gulf countries and fertile east Mediterranean countries. However, the social challenges, the urgent need to create employment, the need for infrastructure, housing and various utilities are common.

Approximately 100 GW were needed to be installed between 2006 and 2015 in the region, at a cost of around $20 billion per annum. There are alternatives to building new power plants, particularly when decision-makers believe in the cyclical nature of costs and can delay the investment decision. These alternatives include demand-side techniques or supply-side strategies. These techniques allow for limited capacity gains in comparison to a large power plant, and assume a certain level of sophistication of the market and additional investments to be spent. However, it is important for all countries to implement efficiency programs and techniques to take advantage of the "hidden generation", to unlock the hidden Gigawatts that these programs yield. More importantly, these programs will contribute to the development of a sound regulatory framework and of a higher knowledge of the electricity system, and will send the right signal to the customer on the value of energy.

These alternatives will not be sufficient to cover the large needs of these countries. In this context, like in other regions, growth in electricity demand, the sense of urgency and the inability of the governments to respond efficiently, played a major role in urging these countries to restructure their electricity sectors. However, the often abusively used binary relation between oil prices and reforms (low oil prices encourage reforms, high oil prices freeze them) proved not to be valid several times in the post-2000 period.

So what were the main factors which really spurred reforms in the region? In addition to the classical drivers often associated with reform movements (poor performance of the State-owned/run electricity sector, inability of the sector to self-finance expenditures), there are three main dynamics which appear dominant: the ambition/temptation to capture revenue from privatization of assets, the urgency to increase performance of the sector and the appeal to align a call for political reform with some kind of economic reform.

The experiences of reform of the different countries in the region are very enlightening in this regard. The behaviour(s) of the State is critical as a driving force. Conventional wisdom would assume that countries would seek to capture a new form of rent. Oil exporting countries would be particularly impelled to do so in an environment of low oil prices and decreasing oil revenue. Non-oil exporting countries would be tempted by these "exceptional receipts" to fill in budget deficits. However, privatizing without restructuring might not be as lucrative as originally thought and there is a limit to what a country can privatize in quantity and in scope. The increasing needs of the population (for basic utilities including water and power) spurred governments to rationalize the management of their budget. The "conditional" allocation of national budgets assumes an implicit ranking of the State's tasks, in order to define which ones will be "privatized". In the long-run, the behaviour of the State vis-à-vis privatization, as a preamble for electricity reform, becomes more dictated by strategic economic orientations rather than opportunistic short-term budgetary considerations.

On performance, accelerated investments and high demand growth rates created or displayed inefficiencies in all parts of the sector. Conversely, it appears that major blackouts—the "ultimate inefficiency"—cannot be necessarily explained by the perceived "weaknesses" of monopoly structures (general lack of reliability, cost reduction mechanism, efficiency incentive…). Actually, the productivity of state-owned utilities in the region is comparable to that of a western vertically-integrated monopoly.

Finally, in some cases, electricity-sector reforms were driven by a general movement of economic reform, which was in turn just substituting to a call for political reform. Then, the decision-makers face several dilemmas. The reform is expected to continue to serve the "interests" of the local population, even if utilities are expected to be "treated as commercial commodities, rather than social benefits". Sometimes, energy and electricity reforms in the region are the heirs of defence policies, with the primary objective to attract foreign interests in the region to guarantee its defence. Therefore, the close involvement of the highest levels of the government, in addition to the interior or defence minister sometimes, can facilitate electricity reform programs but can also complicate them. The third dilemma facing the governments in the region is the need to diversify the economy because of the depletion of fossil fuels, or their scarcity. Decision-makers were hoping to achieve efficiency and fuel gains, in comparison to assets used to provide public services, by introducing the private sector in utilities.

Therefore, up until the middle of this decade, most countries in the region were restructuring their power and water sectors. By 2006, there were only four countries which were completely closed to IPPs for diverse, mainly political, reasons: Libya, Syria, Kuwait and Iraq. Today, with the exception of Kuwait, any major electricity reform is highly challenging in the three remaining countries, and is even probably bound to fail in the existing complex political, institutional and industrial set-ups.

2.5 Summary

The organization of the sector in the different countries of the MENA region has been changing over time, following a series of historical, economic and political changes. The history of electricity in the region dates back to the end of the nineteenth century. MENA countries share several common features (exercise of political power, implicit or explicit contracts between social structures and political leaderships, cultural and religious links. . .etc) which impacted the governance and the organisation of the electricity sector, a strategic and ubiquitous sector. However, these countries can widely differ in terms of economic and human development standards, sources of revenue, private sector development, credit ratings, to cite a few. It should not be surprising to find in this region the world's highest electricity demand growth rates and the highest consumption per capita. Obviously, there is a wide discrepancy between highly populated North African countries, hydrocarbon-rich gulf countries and fertile east Mediterranean countries, but the social challenges, the urgent need to create employment, the need for infrastructure, housing and various utilities are common.

This region is unique. Indeed, high demographic growth, rapid urbanisation and accelerated industrialization are some of the first limits faced when attempting to apply generic models to the MENA region, either econometric models to forecast future investment needs or organisational models which would drive the institutional reforms. Economic growth in the region, conventionally used as a key variable, is "distorted" by international oil price movements and then by the contribution of the hydrocarbons' sector growth to the economy, even in non-oil producing countries. The presentation of the first three experiences of electric reforms in the region (Morocco, UAE and Oman) provides interesting insights on the peculiarities of this region. Morocco attempted to introduce a hybrid market, with limited success, given the uncertainty over prices in the yet-to-be-created deregulated market and over access rules to that market. In the UAE, Abu Dhabi steadily followed a text-book Single Buyer Model and created the Abu Dhabi Water and Electricity Company (ADWEC) in charge of buying all the electricity and water produced, with a limit on foreign ownership of 40% on all Brownfield and Greenfield IWPPs. By contract, Dubai faced strong resistance to electricity reforms. Oman is probably the most advanced in the region as all generation ventures (except the Rural Areas Elec. Company) are 100% privately owned, with an obligation to offer at least 35% of shares on the Muscat Securities Market.

2.5 Summary

The [illegible] of the [illegible] in the different [illegible] of the MENA region has [illegible] in this region [illegible]

[illegible] [illegible]

Chapter 3
Three Main Drivers of Electricity-Sector Reforms

Abstract Electricity-sector reforms started in the mid-1990s in the region, following the "second" global wave of electric reforms. Oman and Morocco led the way in the early 1990s, and they were followed by Abu Dhabi and the majority of other MENA countries. It is conventionally admitted that, in the region, oil prices are an important factor influencing economic reforms. This was probably true during the last two decades of the last century but not afterwards. The need to build consensus among decision-makers guided electricity reforms, particularly in gulf countries. In other countries where interests are concentrated but not mutualised (North Africa), a general suspicion towards anything related to the introduction of the private sector would hamper the process of reform or privatization. The productivity of most vertically integrated monopolies in the region is comparable to that of a western vertically-integrated monopoly. However, the limited financial capability of some could be problematic and would require sovereign guarantee.

Keywords Determinism of oil prices · Consensus · Productivity · Sovereign guarantee

Electricity-sector reforms started in the mid-1990s in the region, following the "second" global wave of electricity reforms which occurred in Latin America and Eastern Europe. Oman and Morocco led the way in the early 1990s, and they were soon followed by Abu Dhabi and the majority of other MENA countries. As in other regions around the world, several factors played a major role in urging the governments of the aforementioned countries' to restructure their electricity sectors: growth in electricity demand, the sense of urgency and the inability of the governments to respond efficiently, to cite a few. Several factors created the need for reforms.

L. Benali, *Electricity-sector Reforms in the MENA Region*, Perspectives on Development in the Middle East and North Africa (MENA) Region,
https://doi.org/10.1007/978-3-319-96268-9_3

3.1 Testing the Limits of the "Oil Price Theory"

It is conventionally admitted that, in the region, oil prices are an important factor influencing economic reforms. Utilities' analysts have often linked the (short) history of introduction of Independent Power Projects/Producers (IPPs) and the acceptance of the contractual schemes enabling the relinquishing of ownership momentarily or indefinitely,[1] to the fluctuation of oil prices and the state of the governmental budgets. This was probably true during the last two decades of the last century but not afterwards. The general perception is that periods of low oil prices create important deficits in the budgets of oil-exporting countries, urging them to encourage the involvement of private investors (mostly foreign ones) in infrastructure projects, including electricity. Conversely, periods of high oil prices were generally blamed for the freezing of reform programs officially or unofficially.

This might have been true in the last three decades of the last century, but this general statement proved wrong several times in the post-2000 period.[2] The period with increasing oil prices post-2002 witnessed the most efforts of reforms in the region as Fig. 3.1 shows. In fact, Middle Eastern countries benefited from an aberration which, if sustained, could have translated into serious prospects of growth: high oil prices with no stagflation in the economies of the major consumers of their hydrocarbons, and in their own economies. However, there were also serious challenges: several MENA economies were still not diversified, and labour forces were growing in excess of 4% a year, requiring an economic growth exceeding 5–7% a year.

The size and activity of the private sector is very variable depending on countries. In Saudi Arabia, the growth of the government share in the non-oil sector outpaced that of the private sector during the 1990s, and accelerated in 1995–1997, a period supposedly favourable for private investments. This trend decelerated after 1997. The private sector share of non-oil GDP remained at 60–70% during most of the two decades after 1997.[3] What is more, a large share of the private sector growth has been overwhelmingly driven by government spending.

Meanwhile, the global economy has proved its resilience to higher oil prices during the first decade of this century. It appeared that it was in a better shape to deal with oil price volatility, particularly when compared to the 1973/1974 and 1979/1980 periods.

---

[1]Boos schemes [Build-Own-Operate (BOO), Build-Operate-Transfer (BOT), Build-Own-Operate-Transfer (BOOT)].

[2]"With two decades of anemic growth following the collapse in oil prices in the mid-1980, most of the economies in MENA had initiated some program of reform to transition to more open, private-sector oriented growth. Resource poor economies like Egypt, Jordan, Morocco, and Tunisia embarked on reform earlier (beginning in the mid-1980s), but the sustained lack of growth inspired almost all countries in MENA to, at least notionally, embrace the need for reform by 2000. Thus, the present oil boom occurs in a markedly different environment than the previous oil booms. This oil boom happens in the midst of an economic transition". Arab Strategy Forum, talking points of session: Will the New Oil Boom Kill the Reform? December 2006.

[3]Source: SAMA.

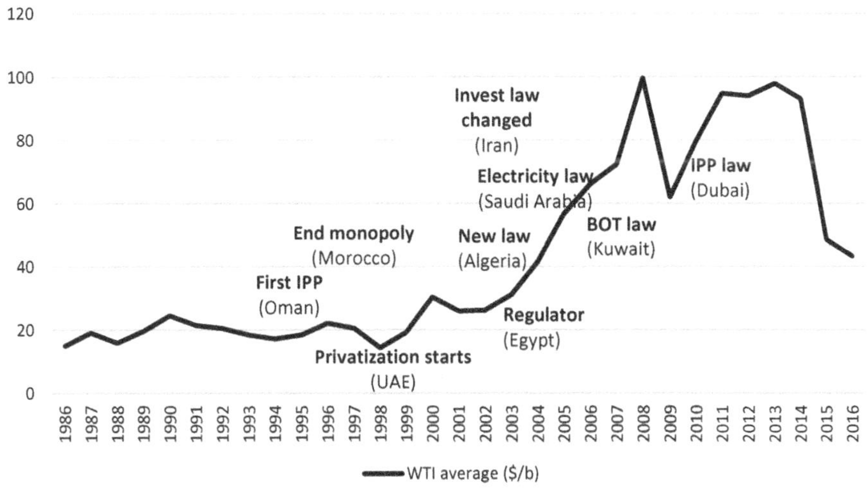

Fig. 3.1 Oil price evolution and electric reforms in MENA. Source: EIA for WTI prices

Post-2000, MENA countries developed several mechanisms to deal with the surge in liquidity associated with high oil prices.[4] For the first time, the momentum to reform did not vanish with increasing oil prices, at least in the electricity sector.

However, there were other threats that oil price volatility would bring as often cautioned by a few oil ministers, including former Minister of oil of Saudi Arabia Ali Naimi. Sustained high oil prices would encourage the investment in Research and Development to develop alternatives to oil, particularly in the transport sector. They could therefore affect the long term demand for oil. Similarly, on the supply side, sustained high oil prices could facilitate the development of marginal expensive oil projects and, as the world would discover a few years later, the development of shale oil with the combination of horizontal drilling and hydraulic fracturing.

Following the US subprime crisis of 2007 and the subsequent global financial crisis of 2008, governments have in general responded by macro-economic measures to keep inflation under control while testing the global economy's resilience to higher oil prices. The US Federal Reserve increased interest rates ten times in 2005 and 2006, and other central banks have followed suit. High oil prices enabled a great bubble of investments in alternatives, including renewable energy, contributed to the revival of clean coal and nuclear power and made the development of costly and marginal upstream resources more economically justified (oil sands, sour gas, offshore Deepwater…).

[4]Mohamed A. El Erian, President and CEO of the Harvard Management Company, at a Luncheon Keynote Speech, ABANA (Arab Bankers Association) Annual Conference, Middle East capital markets, 25 April 2006. New York.

Table 3.1 Saudi Arabia's budget balances and inflation rates evolution. Sources: Jadwa. Saudi Arabian Monetary Agency, Ministry of Finance, EIA

	2002	2003	2004	2005	2007	2008	2009
Average WTI ($/b)	26.18	31.08	41.51	56.64	72.34	99.67	61.95
Budget balance (% of GDP)	−2.9	4.5	11.4	18.4	12.3	33.7	−5.6
Inflation (%)	0.2	0.6	0.3	0.7	4.1	7.4	6.2
	2010	2011	2012	2013	2014	2015	2016
Average WTI ($/b)	79.48	94.88	94.05	97.98	93.17	48.66	43.29
Budget balance (% of GDP)	4.4	11.6	13.6	6.4	−2.3	−14.8	−16.8
Inflation (%)	3.8	3.7	2.9	3.5	2.7	2.2	3.5

In the meantime, oil-producing countries' behaviour has changed as well. Fearing the Jakarta syndrome,[5] OPEC has been particularly careful in managing the market and the level of spare oil productive capacity. The general purpose has constantly been to guarantee a "fair" level of oil prices, fair to both the consumers and the producers. In oil-producing countries' budgets, several consecutive years of high oil prices during the last decade had mixed results. The positive fiscal balances translated into improved economic outlooks and massive domestic investments in infrastructure, education and industries. However, they also created higher inflation rates (see example of Saudi Arabia below) making it more difficult to pursue several reforms, including that of retail tariffs and domestic prices (Table 3.1).[6]

It is only in 2016 that Saudi Arabia could implement a first series of comprehensive energy price reforms, including for electricity and water. These schemes have been in the drawing board for many years, and were meant to be the first phase of a gradual move to market-based prices for most energy products.

Therefore, it is over-simplistic to view the history of power deregulation in MENA as a binary function of the oil price evolution. To be sure, the beginning of the 2000s coincided with the California power crisis and power trading scandals which sent western regulators back to the drawing board. This slowed the development of new reforms internationally, while several reforms in the MENA region were barely starting. For example, Percebois (2008) concluded that electricity liberalization in Europe produced two unexpected, or rather undesired, outcomes: the increase in end-user prices and reinforcement of incumbent operators' market power. Even in France, where the liberalisation process has been kept to the

[5]In November 1997, OPEC decided in its meeting in Jakarta to increase production by 10%. Prices collapsed in 1998, in the wake of the Asian economic downturn. In 1998, oil prices went down from $20 to $10 per barrel.

[6]For a more complete analysis of the context and framework for understanding the unprecedented rise in oil prices since 2001, see "Recession shock: the impact of the economic and financial crisis on the oil market", a special report by Cambridge Energy Research Associates to The UK Department of Energy and Climate Change and the Ministry of Petroleum and Mineral Resources of Saudi Arabia. The report was part of the briefing for the London Energy Meeting on December 19, 2008. It explains why economic growth was not derailed sooner.

minimum and most of the electricity insulated from international hydrocarbons prices as based on nuclear and hydroelectricity, prices were affected by German wholesale electricity prices (more liberalised). More generally, the questioning of the fully deregulated model led the few countries in the region who were considering a full reform of their electricity sectors to stand back.

As the world wavered between "too little" and "too much" regulation, in several regions, governments have been winning the battle between free markets and regulation, with public authorities keeping or regaining control of parts of the electricity sector.[7] It was important to maintain hybrid forms of cooperation, involving public, private and multilateral players, particularly in financing schemes. These were justified by the need to fulfil public policy objectives (energy access, affordability…), even as the sector opens up to competition. In European countries, domestic coal purchase agreements or utility ownership of uneconomic coal mines were meant to support the domestic coal mining industry.

In MENA countries, similar objectives would exist, including oil-oriented policies or the support of domestic fuel production. In these cases, fuel allocation policies would push utilities and heavy industries to consume favoured or discounted fuels, crude oil for instance. In the same vein, energy security considerations would require the government to remain involved in the sector. Some fuel sources would be selected at the expense of others. The use of non-competitive procurement procedures or the obligation to use a specific generation technology, even on uneconomic grounds, would be designed to develop a local equipment supplying industry (local content). Employment needs would in a few countries justify over-staffing utilities.

Another policy objective is related to the concern of cyclical deteriorations of reserve margins. Global experiences show a wide diversity in the way deregulated markets have attempted to deal with the issue. Capacity payments are the mechanisms that some electricity markets have decided to implement. In the UK for example, it consists in determining the Loss of Load Probability (LOLP) and the damage such a shortfall would cause to consumers (Value of Lost Load VOLL). Capacity mechanisms are also used in Spain and Argentina. Most MENA countries have decided to bear the cost of comfortable reserve margins themselves, with the possibility of freezing a reform program and turning to fast-track EPC projects in order to get the plants built. This should happen regardless of oil prices position. In MENA, as in other regions in the world, supply interruptions and price volatility are unacceptable to the different stakeholders (government, large customers, population … etc) (Table 3.2).

High oil prices and accelerated economic growth in the region resulted in accelerated investments in the electricity sector, and in inefficiencies. These inefficiencies are particularly flagrant on the demand side. The region records some of the highest electricity consumptions per capita in the world, partly because of consumers'

[7]The Hirschmann dialectic describes this oscillation between the market and the government and the intellectual delusion vis-à-vis Keynesianism that it produced.

Table 3.2 Benefits and risks of various capacity mechanisms

Capacity mechanisms	Benefits	Risks
Vertically integrated monopoly or concentration	Price stability Security of supply Regular investments (if available funding)	Efficiency Cost/quality of service Technology choices
Security of supply commission (New Zealand)	Security of supply	Premium on spot prices
System Operator as guarantor (Colombia, Sweden)	Reliability separate from adequacy	System Operator as market participant by soliciting bids—capacity reservation or direct ownership of reserve capacity
Capacity payments (Argentina, UK)	Security of supply	Overcapacity Gaming
Capacity markets (PJM)	Capacity valued separately	Gaming the system
Renewable subsidies	Investments	Expensive Reliability/adequacy not necessarily improved

behaviour in an environment of highly subsidized tariffs but also on insufficient investments in Demand Side Management's best practices (buildings isolation, air-conditioning in commercial buildings, public lightning…).

In reality, there has been more emphasis on financing supply to avoid shortages than on managing demand and addressing operational deficiencies. Inefficiencies are noticeable on the generation and transmission side, where proper dispatch and system optimization are not totally implemented. Actually, the transmission system could be a real hurdle to reform in an environment of growing capacity and demand. Transmission networks were built to serve a system based on regulated monopolies, less to connect buyers and sellers with an economic incentive to seek the best deal. The creation of a market requires having the technical ability to "send power in different directions, to different customers at different times at different prices in response to different demands".

The growth of industrial auto-generation, driven by ambitious industrialization, created "islands" of electricity generation centres. The captive power plants, associated with heavy industrial plants, were usually tendered as IPPs, and typically more efficient than existing assets. These captive power plants contributed to the general move towards breaking monopolies, on the condition that they are efficiently integrated into the whole system.

In this context, these are all new considerations to the relationship between oil prices and electricity reforms when applied to the region. The dynamic of this relationship is different today than it was during the pre-2000 period. Oil price movements continue to impact electricity reforms but not in the conventional "binary" way (i.e. high oil prices stop reforms, low oil prices encourage them). The economic growth brought by the 2003–2014 period of high oil prices did not discourage the willingness to reform. There has been strong momentum to reform, but the implementation itself faced hurdles. Moreover, most electricity-sector reforms in MENA coincided with the questioning of full deregulation, leading most to halt reforms at the Single Buyer Model.

International experts caution against "one-size-fits-all" models, and recognize the variations in context across countries and timings. Anupama Sen et al. (2016) argue that electricity reform programs, did not always lead to higher efficiency, maximize economic welfare, and transfer surplus to consumers, even in the OECD economies which pioneered the standard model. In an analysis covering 17 non-OECD developing Asian economies spanning 23 years, the authors show a tension between wider economic impacts and welfare impacts for consumers and that institutional factors have influenced the outcomes. The uniform application of the standard model without reference to the heterogeneity of the countries is unlikely to have resulted in originally intended outcomes.

3.2 The Real Driving Forces: Performance, Revenue, Financing, or Reform Calls

A number of authors (Bacon 1995; World Bank 2002) have pointed to a series of factors behind reform movements: poor performance of the state-owned/run electricity sector (costs, access to service, reliability…), inability of the sector to self-finance capital, or even O&M expenditures, the need for the government to capture revenue through privatization of assets or through the change of the tariff structure and the decrease/removal of subsidies.

In the region, the drivers that led the countries to introduce some level of competition include all the factors cited above. The observation of empirical experiences in the region indicates that there are however three main drivers which appear dominant:

- The need to capture revenue from privatization of assets
- The urgency to increase the sector's performance, through the introduction of competition
- The appeal to align a call for political reform with some form of economic reform.

On the Need to Capture Revenue from Privatizations

Privatization is any "transfer of assets from the government to the private sector, could they be issued from a previous nationalization or not". Inevitably, the act of privatizing relates to the identity and role of the State. In MENA, the State has probably "outsourced" part of its functions, providing a public service, but has surely not completely retrieved from the sector.

Previous research (Benali 2002) highlighted the continuity of the function linking the possible organizational designs of the electricity sector with the degree of State intervention in the sector. This enabled us to analyse the different possible organizational designs, to define where the Abu Dhabi electricity and water sectors could be positioned, and to present the characteristics of the various degrees of State interventionism, with a middle way degree of state intervention in the sector, called *Liberal Interventionism* (Fremaux 2002).

The behaviour(s) of the State, as a Group of Economic Orientation, is therefore critical as a driving force. Conventional wisdom would assume that countries would seek to capture a new form of rent. Oil exporting countries would be particularly pressed to do so in an environment of low oil prices and decreasing oil revenue (as it was the case in the pre-2002 period). Non-oil exporting countries would be tempted by these "exceptional receipts" to fill in budget deficits. Here again, this statement might be over simplistic in the sense that it does not capture the complexity of the behaviour(s) of the State. The Iranian example (see Box) shows how privatization is much more complex than the simple selling of assets.

Iran's Privatization Efforts: When the Clergy Comes Into Play

Privatization has been on the agenda since the first post-revolutionary 5-year plan of 1989. However, it has been either fruitless or resulting in the creation of large "institutions" evolving in a "grey zone" between government and private sectors; for two main reasons:

- A strict interpretation of article 44 (amended by two decrees in 2004 and 2005) of the Iranian constitution:

> The economy of the Islamic Republic of Iran is to consist of three sectors: state, cooperative and private, and is to be based on systematic and sound planning. The state sector is to include all large-scale and mother industries, foreign trade, major minerals, banking, insurance, power generation, dams and large-scale irrigation networks, radio and television, post, telegraph and telephone services, aviation, shipping, roads, railroads and the like; all these will be publicly owned and administered by the State. The co-operative sector is to include co-operative companies and enterprises concerned with production and distribution, in urban and rural areas, in accordance with Islamic criteria. The private sector consists of those activities concerned with agriculture, animal husbandry, industry, trade, and services that supplement the economic activities of the state and cooperative sectors. Ownership in each if these three sectors is protected by the laws of the Islamic Republic, in so far as this ownership is in conformity with the other articles of this chapter, does not go beyond the bounds of Islamic law, contributes to the economic growth and progress of the country and does not harm society. The scope of each of these sectors, as well as the regulations and conditions governing their operation, will be specified by law.

- A general suspicion from the private sector towards political uncertainty, non-representation in the country's political bargaining process and the rise of some companies as government's favorites, particularly during contracts' awards.

(continued)

The emergence of political entities affiliated with the business community (Society of Iranian Protectionists) or the issuance by Supreme Leader Khamenei of two decrees (in 2004 and 2005) amending Article 44 were deemed as positive steps of reform. These amendments called for instance for the withdrawal of the State from all areas in which the private sector can participate by 2010 and that government agencies will become regulatory and supervisory bodies. However, a significant proportion of privatized companies landed in a "grey zone" between government and private sectors (religious foundations, pension funds, social security organizations).

Even those countries which focused their budgetary policies on privatization revenues swiftly understood that privatizing assets without first investing effort in restructuring them might not be as lucrative as originally thought. Morocco is a good example in this regard: the country embarked in an ambitious privatization program in the late 1990s, in addition to the reform of its electricity sector, partly to modernize the industrial landscape of its 127 national offices and companies. The 2005 budget benefited from the cession of 16% of Maroc Telecom—12.4 billion dirhams (1.2 billion Euros)—but that was more the exception than the rule. Remaining companies on the list were not expected to be sold at more than 5 billion dirhams on aggregate. This was one of the reasons encouraging the authorities to turn to concessions and encourage public-private partnerships. They drafted a framework Legislation, with application decrees which would take into account sectorial specificities, including for electricity. In 2015, the national utility ONEE was still recording a deficit of 2.3 billion dirhams. A framework agreement with the State over the 2014–2017 period covering capital injection and tariffs increase, and coinciding with fuel costs decrease, was expected to enhance the company's financials.

In general, the increasing needs of the population for basic utilities including water and power spurred governments to rationalize the management of their budget. In parallel to this "rationalizing" mood, the role of the State has also been evolving. National budgets started to be preferentially allocated to activities that are naturally within the responsibility of the State: health, education, defence, foreign policy ... etc. This "conditional" allocation assumes an implicit ranking of the State's tasks, in order to define which sectors will be "privatized".[8] For example, the defence budget in MENA countries (5% of GDP in 2007, down from 6.5% in 2001) is particularly important: important relative to other countries (4% in the US in 2007, 1.5% in Russia, 2.6% in NATO countries) and relative to other key areas in the budgets (5% for construction, 4.3% for education and 5% for health—source: Arab Monetary Fund).[9] For instance, World Bank data estimate military expenditure in Algeria,

[8]Béatrice Hibou (dir.), *La privatisation des Etats*, Paris, Karthala, 1999.

[9]For more details on MENA's defense budgets, refer to Appendix E.

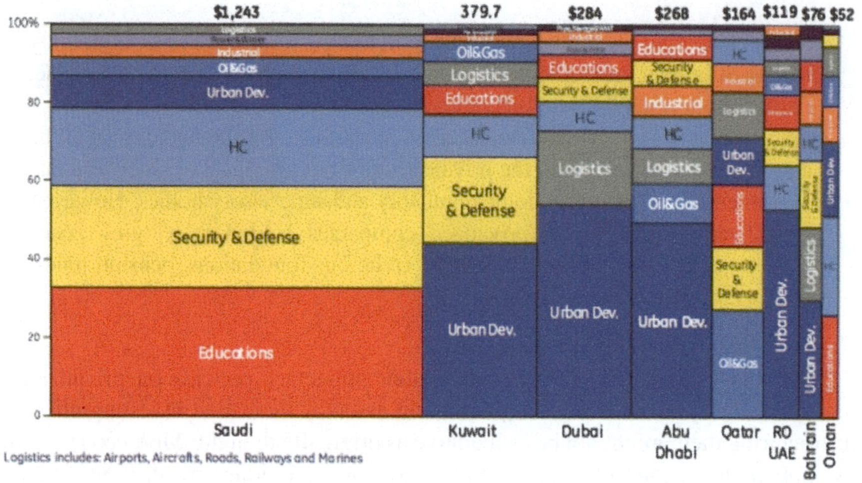

Fig. 3.2 2007–2020 planned spending in GCC countries. Source: General electric, national budgets (2007)

Morocco and Saudi Arabia to range from 3.3 to 9.6% of GDP in 2016 (data unavailable for all countries) (Fig. 3.2).

In selected cases, revenues from privatization lured governments and encouraged them to open their electricity sectors to foreign and private capital. In those cases, the "reform" occurred amid a wave of privatizations impacting several sectors of the economy. However, in the long-run, the complex behaviour of the State vis-à-vis privatization, as a preamble for electricity-sector reform, becomes more dictated by strategic economic orientations rather than opportunistic short-term budgetary considerations.

On the Urgency to Increase Performance of the Sector

As mentioned previously, high oil prices and accelerated economic growth in the region resulted also in accelerated investments in the electricity sector, and therefore in inefficiencies in all parts of the sector. One illustration is when massive investments in generation were not accompanied with associated investments in transmission. In the meantime, gas and electricity demand have been registering high growth rates, constraining the transmission and distribution systems and displaying some of the fragilities of the energy chain. For example, gas shortages, particularly during the summer electricity peaks, lead some utilities in the region to run diesel generators and Combined Cycle Gas Turbines (CCGTs) on diesel.[10] The phenomenon is not

[10]See Barrett and Benali (2007).

Table 3.3 Productivity and performance of selected national utilities. Data for 2016, source AUPTDE

	Productivity (MWh/employee)	TWh produced
Jordan	2355	19
Tunisia	2080	18
Algeria	2157	66
Saudi Arabia	9002	345
Syria	512	19
Oman	10034	28
Kuwait	3522	70
Lebanon	3992	11
Libya	802	36
Egypt	1064	186
Morocco	3186	30

new but it is becoming more important in the lack of major gas developments in the region. In the worst cases, the combination of various inefficiencies and of the systems' constraints results in major blackouts.[11]

It is interesting to note that the electricity sector was not reformed in most countries where major blackouts occurred in recent years. Can this be solely explained by the perceived "weaknesses" of monopoly structures (general lack of reliability, cost reduction mechanism or efficiency incentives)? Not necessarily. The following table shows the disparities (Libya vs. Iran for example) between a few selected countries, using the ratio of productivity (GWh generated per employee) as one indicator for performance. Furthermore, with the exception of Israel and Iran to a certain extent, none of these electric systems have experienced the introduction of independent electricity generation or any other form of competition. Yet, barring the exception of Libya and Yemen, their productivity is comparable to the productivity of a western vertically-integrated monopoly. In Libya and Yemen, the official number of employees is higher than the actual number, therefore the productivity ratio is distorted (Table 3.3).[12]

[11]Between 2004 and 2008, there were five major blackouts reported in the region: in Jordan (the whole country in August 2004), in Kuwait (the whole country in November 2004 and summer 2006), in Bahrain (the whole country in August 2004) and in Dubai (the whole emirate in June 2005). Of course, Iraq has been experiencing recurrent power shortages.

[12]After the lifting of sanctions on Libya in 2003, the government embarked in a program of national reform. This program was meant to ineluctably impact the electricity for two main reasons. First, the development of this sector is a prerequisite for the development of any other sector, which in itself in an incentive to seek a higher level of performance. Second, the authorities are increasingly concerned about the high level of subsidies injected in the sector, which would decrease if there is a better performance, and about the inability of the sector to finance its operations and invest in new projects. For more details on this period of Libya's economic history, see the National Economy Strategy report developed by Monitor and CERA.

Obviously, these metrics are not sufficient to judge the performance of these utilities. For example, IEC's financial situation has been dire as it accrued substantial debts and is unable to make payments. It struggled to find the funding for the state-backed emergency generation program, but has been reportedly continuously opposed to private generation. In an effort to push forward with greater private power generation in the country, the Finance Ministry, in an agreement with the Ministry of Infrastructure, pledged in July 2009 to guarantee payments to IPPs,[13] if IEC fails to make payments.[14] IEC will still be liable to make payments, and can be sued for not making those payments, although private producers now have the option to bypass it and refer to the Finance Ministry directly.[15]

The difficult financial situation of most national utilities in the region hampers their capability to raise funds to invest in needed projects. This ultimately creates performance problems for the whole electric system when the utility is not backed by a financially strong State who can either provide funds directly or a strong sovereign guarantee for investors. This statement is less relevant for Gulf countries' utilities.[16]

On the other hand, as several authorities understood that privatizing assets without restructuring them might not be as lucrative as originally thought, they also realized that reforming the sector without the appropriate technical and contractual framework might create congestion, reliability and coordination problems. The California crisis of 2000–2001 was a major wake-up call on the limits of partial deregulation (market manipulations, unmanaged increase in wholesale prices with capped retail prices), but a few similar problems occurred in other parts of the world and the MENA region as well even if most reforms were light-handed. In the case of Bahrain, where Aluminium Bahrain (ALBA) stands as an independent power generator, the privatization of the electricity sector was championed by the Crown Prince and Chairman of the Economic Development Board, Sheikh Salman bin Hamad al Khalifa, and facilitated by the Prime Minister Sheikh Khalifa bin Salman al Khalifa. The political will was strong, despite expected resistance from the Ministry of Electricity and Water.[17] But in August 2004 (with temperatures around 50°C), the country experienced a major black-out because of a massive power surge,

[13]Israel targets to have 20% of power needs covered by IPPs in 2020, up from less than 5% today.

[14]The rather unusual agreement will only apply to producers who secure financing over the following 30 months. It does not derogate the IEC's obligation to make payments approved by the Public Utilities Authority.

[15]Tariffs to private generators of 8%/kWh were agreed in November 2008 by the previous government.

[16]In Aujourd'hui le Maroc (4 January 2008), a Moroccan economic daily, the Director General of the Moroccan National utility ONE stated that the financial situation of ONE in 2007, even if improving in comparison to 2006, remained fragile. The difficult international energy context and the freeze of electricity tariffs contributed to its financial difficulties. Increases in fuel costs are reportedly bore exclusively by ONE (neither by IPPs nor by distribution companies). ONE takes care of the public service (rural electrification, security reserves), has an important debt and some distribution companies still accrue liabilities.

[17]In June 2003, Ernst & Young were hired as consultants to advise on the privatization program.

which was reportedly blamed on ALBA.[18] The blackout unveiled the weaknesses of the country's electric system, spurring the Ministry to ask for the equivalent of $1.9 billion to upgrade facilities and infrastructure. Actually, even before the 2004 blackout, a Committee was formed to look at the reasons of more limited disruptions to electricity supplies.

Another area where performance improvement is needed is electricity delivery. 15–25% of system's losses in MENA countries were assumed earlier. In this regard, the top performers are not the ones one would think. According to the IEA, Iraq and Syria were registering, until 2003, 6–7% of system's losses. Smaller countries like Lebanon and Tunisia recorded losses in the same order of magnitude. But for most, losses include non-technical losses and unaccounted for consumption. A few have unbundled and privatized the distribution part of the chain, with the hope that the private sector will be better positioned than the public authorities to reduce non-technical losses. Less than five countries in the region have privatized the distribution sector, or have been seriously considering doing so.

In a nutshell, the perceived weaknesses of monopoly structures are not sufficient to explain the problems of performance of the electricity sector in some MENA countries. For example, the productivity of most vertically integrated monopolies—or quasi-monopolies—in the region is comparable to that of a western vertically-integrated monopoly. However, the limited financial capability of some could be problematic. In deregulating countries, authorities realized that reforming the sector without the appropriate technical and contractual framework might create congestion, coordination and reliability problems, even in the smallest and most manageable systems.

On the Appeal to Align a Call for Political Reform with Economic Reform

The rise of political Islam, the decline of the Soviet Union and the increasing calls for political reforms and regime changes in the region led governments to consider a range of economic reform programs, well before the Arab Spring. To avoid the loss of political control, some government officials sought to emulate the Chinese model of reform, which promoted economic reform and modernization in many sectors while the Chinese Communist Party retained complete political control.[19]

In Middle Eastern countries, particularly hydrocarbon-exporting ones, utilities (power and water) were treated as "just another set of social benefits". There is a wide literature on the concepts of "democracy of bread" and "rent-redistribution"

[18]ALBA was established in 1968. It is 77% government owned (77% Bahrain Mumtalakat Holding, 20% SABIC and 3% Breton investments).

[19]See S.L. SHIRK, The Political logic of economic reform in China, University of California Press, Berkeley, 1993.

applied to the region (Sadiki 1997; Beblawi and Luciani 1987). In summary, these social benefits are part of a social contract between the ruling class—or family—and the people. The latter will not question the power of the former as long as the rent from hydrocarbons' exports is *equitably* distributed and basic utilities are provided for free, or for a very low tariff. Fattouh et al. (2016) argue that, contrary to conventional wisdom, the implicit social contract in the GCC has proved to be elastic and sufficiently malleable to accommodate energy price changes.

Against this background, decision-makers were facing their first dilemma: the reform is expected to continue to serve the "interests" of the local population, even if utilities are expected to be "treated as commercial commodities, rather than social benefits".[20] Providing utilities at below-market prices or for free was contributing to runaway energy demand in several areas, further exacerbating the potential for shortages.

The second dilemma lies in the defence strategy of these countries. Energy and electricity reforms in the region are at times the heirs of defence policies, particularly after the first Gulf War in 1990–1991. Like offset programs or the Jebel Ali zone, the electric-sector reforms' preliminary objective would be to attract foreign interests in the region to guarantee its defence. This theory is partly supported by the close involvement of the highest levels of the State, and the interior or defence minister, in electricity reform programs.

The third dilemma facing the governments in the region is the depletion of fossil fuels, or their increased scarcity. Dubai or Bahrain were among the first ones to diversify their economies because their oil and gas production rates were decreasing. They pursued low-energy intensive/high employment strategies (financial services, tourism) to achieve varying levels of economic diversification. Others had to follow suit, because their domestic fuel demand started to exceed their domestic supply and because regional resources were becoming scarce and more expensive. Decision-makers were hoping to achieve improvements in efficiency and fuel gains (compared to the performance of historic assets used to provide public services), by introducing the private sector in utilities.

However, there were several occasions where these grand political and macro-economic considerations were overshadowed by mere financial interests. Indeed, rent distribution is the nexus of the tacit contract linking the population to the ruler. In several MENA countries, where major financial interests are concentrated within a relatively small group of individuals or families, IPPs may appear lucrative in the sense that potentially large rent could be collected by developers from selling output at fixed prices. Board representation in major energy and industrial companies could resemble a musical chairs game between a handful of key individuals. This could be justifiable given the size of the population in some countries.[21] In other regions in the world, private power projects were associated with corruption scandals.[22] In MENA,

[20]Allen Conroy, Deputy Managing Director of Abu Dhabi's Umm al Nar Power Company.

[21]For example, 2.7 million inhabitants in Oman, and 1.6 million in Qatar, including expatriates.

[22]See for example Keefer, P. (2002), "the political economy of corruption in Indonesia", World Bank, Washington, DC.

tendering and bidding processes have been generally open and transparent. The need to build consensus among decision-makers have mostly guided electricity reforms, particularly in gulf countries. However, this consensus was in turn sometimes led by financial and industrial interests of powerful individuals.

In other countries in the region, where interests are concentrated but not mutualised (North Africa), a general suspicion towards anything related to the introduction of the private sector would hamper the process of reform or privatization.[23]

This section attempted to summarize the main reasons why, by 2000 and up until the middle of the last decade, most countries in the region were restructuring their power and water sectors. The following table gives a view on the various stages of reform in the region in 2000 (Table 3.4).[24]

By 2006, there were only four countries which were completely closed to Independent Power Projects for various, mainly political, reasons: Libya, Syria, Kuwait and Iraq.

Iraq and Libya explored the reform of their electricity sectors at various moments of their history. However, the two countries have been experiencing long years of sanctions and embargoes, in addition to three wars in the case of Iraq. Today, with sanctions being lifted on both and given the urgency to rebuild or expand the power sector, Tripoli and Baghdad have considered the introduction of IPPs but the complexity of the organizational structures, the legacy of centrally-planned economies and the dispersion of the decision-makers between various priorities made it extremely difficult so far to launch a reform program in these two countries.

Kuwait's electricity sector has been gripped by political wrangling between the country's strong Parliament and the Ministry of Electricity and Water (MEWA). Similar problems have plagued oil sector projects. Even if the two sides agree that

---

[23]Similar trends occurred in other parts of the world. For example, "surveys show that a substantial share of the Honduran population does not see privatization warmly . . . this is partly explained by the suspicion that questionable deals were entered into" Walker and Benavides (2002).

[24]The European Bank for Reconstruction and Development defined the levels of sector transition according to more complex criteria: **level 1:** Power sector operated as a government department; political interference in running the industry; few commercial freedoms or pressures; average prices below costs, with external and implicit subsidy and cross subsidy; very little institutional reform, with monolithic structure with no separation of different parts of the business. **Level 2:** Power company distanced from government, e.g. joint-stock company, although still political interference; some attempts to harden budget constraint but management incentives for efficient performance weak; some degree of subsidy and cross subsidy; little institutional reform; monolithic structure with no separation of different parts of the business; minimal, if any, private-sector investment. **Level 3:** Law passed accounting for full-scale restructuring of the industry, including vertical unbundling through accounting separation, setting up a regulator; some tariff reform and improvements in revenue collection; some private involvement. **Level 4:** Law for industry restructuring passed with unbundling of the industry and regulator set up with rules for cost effective tariff-setting formulated and implemented; arrangements for network access (negotiated access, single-buyer model) developed; substantial private sector involvement in distribution and/or generation. **Level 4+:** Business vertically unbundled; independent regulator with full power to set cost-reflective effective tariffs; large-scale private-sector involvement; institutional development covering arrangements for network access and full competition in generation.

Table 3.4 The state of reforms in MENA in 2000

	Relative openness to IPPS	Effective move towards deregulation	Single Buyer Model	State monopoly (or −2% of private generation)
Morocco	Open	No	X	
Algeria	Semi-open	No		X
Tunisia	Open	No	X	
Libya	Closed	No		X
Egypt	Open	No		X
Israel/ Palestine	Closed	No		X
Jordan	Open	No	X	
Lebanon	Semi-open	No		X
Syria	Closed	No		X
Saudi Arabia	Closed	No		X
Yemen	Closed	No		X
Oman	Open	No	X	
UAE	Open	No	X	
Qatar	Semi-open	No		X
Bahrain	Open	No	X	
Kuwait	Closed	No		X
Iran	Semi-open	No		X
Iraq	Semi-open	No		X

urgent investments are needed to stave off emergencies and blackouts, the Parliament rejected various MEWA-sponsored bills. Some of these bills were related to the purchasing process of turbines where the Centrals Tenders Committee of the Parliament has to be involved. This constant struggle hampered any advanced discussion on reforming the electricity sector until 2008.

In Syria, when President Bashar al Assad came to power in 2000, he started by abolishing four ministries including that of planning. The abolition of the Ministry of Planning signalled the abandonment of central planning. Thereafter, the Syrian Government finalized a draft Economic Reform Program (ERP). The program sought to overhaul the economy and has set a number of objectives to be achieved within a period of 5 years. These included financial reforms[25] and reform of the judicial system. This wide plan of economic reforms raised the hopes that the electricity sector might follow suit, but progress toward implementing urgently-needed projects has been slowed by a lack of investment capital, mainly because

[25] like freeing interest rates, free transfer of funds, unification of exchange rates, full convertibility of the Syrian Pound, independence of Central Bank, revision of income tax, revision of trade laws, modification of custom tariffs, setting up of a Stock Exchange.

of the sanctions[26] discouraging international banks and western players to be involved. The state-run Public Establishment for Electricity Generation and Transmission (PEEGT) controls all aspects of generation and transmission, while the Public Establishment for Distribution and Exploitation of Electrical Energy (PEDEEE) is responsible for sales and distribution. In order to attract the necessary capital, Syria has tried to open the sector to IPPs, although its similar reform efforts in the early 1990s failed. A long delayed project for a 450-MW single cycle power plant in Sweidieh, signed with Iran's Azrab Energy Industries Development Company to build and operate, has not materialized. The plant was supposed to start production in mid-2009. Other attempts to attract foreign Joint Ventures (Iberdrola and Alstom Poland, Siemens and Koch) failed to be launched as IPPs. The other challenge that the country has been facing is its dwindling oil production and its limited indigenous gas supplies.[27] The fuel challenge made it more difficult to attract the right players to the country's electricity sector.

Today, with conflicts raging in each of these three countries, it is difficult to imagine how the electricity sector could be dramatically reformed in each of Syria, Libya and Iraq without a major change in the political landscape of these countries, despite the dire need to improve the performance of the sector. Most have tried to launch economic reforms at the expense of political reforms. Some are in desperate need of revenues from privatizations or of the involvement of the private sector. However, given the different political struggles in these three countries, any major electricity-sector reform is bound to fail in the existing fragile political and industrial set-up of these countries.

Kuwait created in 2008 the Partnerships Technical Bureau to engage private investments, and decreed that power plant projects will be developed through public-private partneships (PPP) projects as BOOTs (Build, Own, Operate, Transfer). Kuwait's first IWPP Az-Zour phase 1 reached financial close in 2014, with GDF Suez, Sumitomo and al Sager group as the owners. All IPP/IWPP projects sign power and water purchase agreements (PPAs) with the single buyer, the Ministry of Electricity and Water (Table 3.5).

In Israel and Iran, less than 2% of the electricity was generated by private players.

In Israel, less than 1% of the installed capacity is owned by IPPs. The Electricity Sector Law was enacted in 1996, replacing IEC's 70-year concession to generate, supply and distribute electric power. A new Law was passed in 2004 where the

[26]The USA PATRIOT Act (the Uniting and Strengthening America by Providing Appropriate Tools Required to Intercept and Obstruct Terrorism Act) of 2001 and the Syria Accountability Act of 2003.

[27]Syria oil production has been in constant decline since its peak of 640,000 barrels per day in 1996, due to technical deficiency, low investments and steady depletion. Crude production was estimated at 400,000 barrels per day in 2005.

Table 3.5 The state of reforms in the region in 2006

	Relative openness to IPPS	Effective move towards deregulation	Single Buyer Model	State monopoly (or –2% of private generation)
Morocco	Open	X		
Algeria	Open	X		
Tunisia	Open		X	
Libya	Semi-open			X
Egypt	Open		X	
Israel/ Palestine	Open			X
Jordan	Open		X	
Lebanon	Semi-open	X		
Syria	Closed			X
Saudi Arabia	Open	X		
Yemen	Open		X	
Oman	Open		X	
UAE	Open		X	
Qatar	Open		X	
Bahrain	Open		X	
Kuwait	Closed			X
Iran	Semi-open			X
Iraq	Semi-open			X

Public Electricity Utility Authority sets criteria for IPPs.[28] One of the stated goals of the Electricity Law is to set a target of increasing the generation of electricity by independent power producers to 20%. The reform of the Israeli electricity sector has been suffering from the legacy of a rigid system, coupled with IEC's limited financial performance, particularly as the company has to finance operations and investments from internal cash flow and loans from various sources.

[28]In order to try to minimize production costs and encourage the sale of electric power by IPPs, the Minister of National Infrastructures enacted the Electricity Sector (Conventional Independent Producer) Regulations-5765–2004. These regulations determined a mechanism where an IPP could require IEC to purchase available production capacity for a payment equal to 80% of the normative cost of the production unit (an objective cost to be determined by the Electricity Authority) for each MW of available supply capacity that it provides IEC and 100% of the fixed operating and maintenance costs for each MW of available capacity. This mechanism serves as a sort of "safety net" to guarantee full payment of the debt in case the producer does not sell the station's full capacity. The regulations add that where the producer provides available production capacity to IEC, the producer may be obligated to provide electricity to the national grid at a fixed rate to be determined by the Electricity Authority (may reflect the production cost). The regulations also provides for the stability of production components or normative costs in IPP transactions, except for the regular updates as determined by the Electricity Authority.

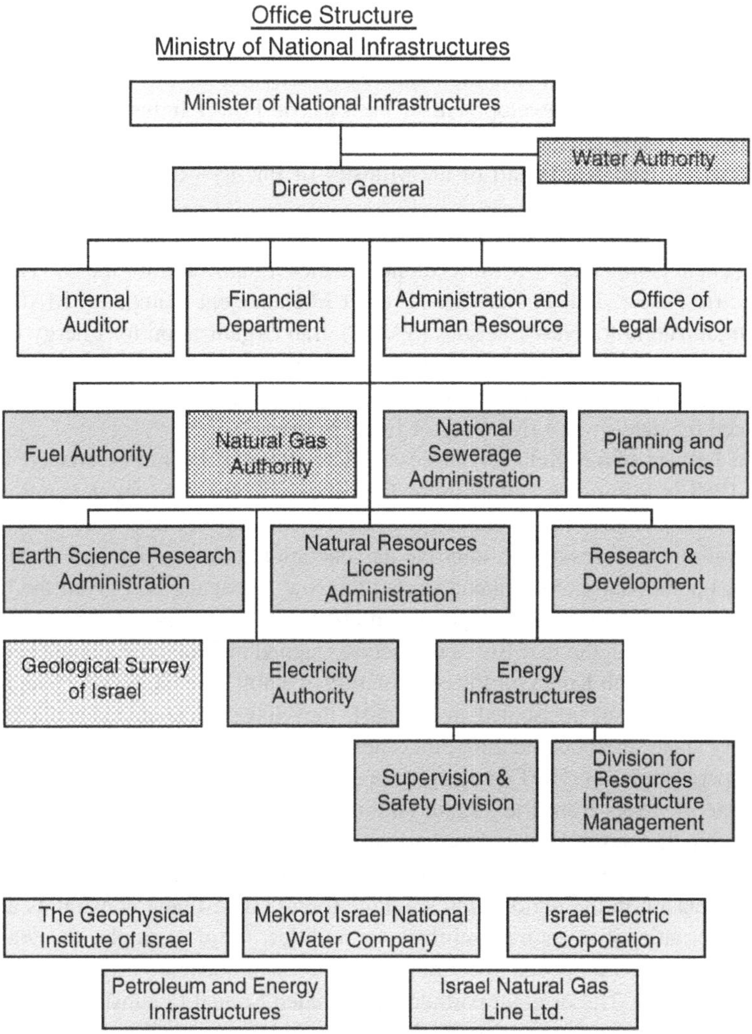

Fig. 3.3 Energy sector structure in Israel. Source: Ministry of National Infrastructures

The power is dispersed between the ministry, the various government bodies, the utilities and the private companies' lobbying. The authorities attempted to break the Single Buyer Model with no critical mass to support the move. This resulted in a hectic deregulation with a non-optimized regulated market and an uncontrolled parallel deregulated market. From a private developer's perspective, the Israeli market's main weakness has been its vulnerability in a politically difficult region. The difficult gas supply situation has been improved with the latest large offshore gas discoveries. Its main strength is its relatively small size offering a wide range of opportunities (Fig. 3.3).

In Iran, the structure of the electricity sector is even more complex. Tavanir, the Iran Power Generation, Transmission and Distribution Power Company, was established in 1979 to "faster the supervisory activities of the government in the field of operation and development of the Electric Power Industry of the country within the scope of policies issued by the Ministry of Energy". It does not only have a supervisory role—on behalf of the Ministry of Energy—and can be involved in certain execution activities.

It leads 16 regional electric subsidiaries, 42 distribution companies, 27 generation management companies.[29] It works alongside the Organization for the Development of Electric Power, Iran Power Plant Project Management Company (MAPNA),[30] Iran Organization for New Energies (SANA), Iran Organization for Energy Productivity (SABA) and Iran Power Plant Repairs Company, which are all its subsidiaries in theory. Only 2% of the country's electricity is generated by IPPs, all the rest being produced by state-owned (Ministry of Energy) assets.

The Iranian power industry successfully weathered the hiatus created by the 1980–1988 Iran-Iraq war and its immediate aftermath as electricity demand growth in the 1990s has been lower than expected. The country was then able to introduce conservation measures and to improve its generation, transmission and distribution systems operationally. As demand started to grow fast again, and given the lack of foreign exchange to finance capacity additions, the government turned to BOT schemes. However, the first BOT contract was signed in 2009 with private investors from Iran and South Korea for the construction of a 500 MW CCGT. Although BOT contracts have been negotiated in the past, they have never actually reached the signatory stage. In 2000 and 2001, the state electricity agency Tavanir awarded five build-operate-transfer (BOT) projects (representing a total of 5000 MW) to a number of private investors from Iran, Saudi Arabia, and Europe. Unfortunately, this coincided with the onset of the downturn in global power investment and finance. Domestic political squabbles over foreign investment are combined with the rising political tension in the region, Iran's unique foreign investment constraints and the tough international investment climate for utilities.[31] For example, the 900 MW Parehsar plant was supposed to be Iran's first IPP, with financing originally expected to close in 2003. The original partnership included Sondel (a subsidiary of Edison SpA), DSD (Germany) and MAPNA, but negotiations with Tavanir derailed the process. The Jalal project, another BOT scheme considered around the same time, was effectively halted as its primary sponsor, ABB, moved to disinvest from all its international assets.

[29]The Generation Management companies maintain and operate the 53 state-owned thermal power plants of the country. In addition to these thermal power plants, the country accounts 17 hydroelectric power plants.

[30]Development of thermal power plants—through local or foreign contractors—is the main duty of MAPNA.

[31]The ongoing US economic sanctions against Iran prohibits US companies or banks from participating in Iranian IPP projects.

The Iranian electricity sector suffers from a few shortcomings:

- First, Tavanir plays different roles. It is a regulator, a planner, a trader and an investor in the country's electricity sector, just to cite a few roles. Its list of duties include regulatory (preparation and proposal of electricity Tariff to the Ministry of Energy or of necessary proposals in the field of strategies, policies, long term/ short term planning of the Electric Power Industry and presenting them to the Ministry of Energy for approval) or investments aspects (Investment on Generation, Transmission and Distribution facilities, Obtaining any kind of loan, financial facilities from local or foreign resources, bonds and local participation in investment and advance selling of right of connections and electricity or other methods of providing financial resources).
- Second, supporting services (repair of power plants, network, supply of goods. . .) are carried out by Tamirat Niro Company and some semi-governmental companies gathered under the umbrella of a holding called Satkab company. The nature of this holding, its relation to the ministry and its responsibilities are unclear. But in any case, even supporting services fall within the responsibility of a quasi-monopolistic limited circle of companies.
- In the same vein, constraints on foreign investments in the energy sector and the Iranian general nationalistic stance have resulted in the promotion of quasi-government companies. It also helped flourishing a mix of local contractors that are increasingly involved in the energy sector. These contractors tend to fall into three categories:

  - Long-standing industrial companies and EPC contractors such as Sadra Group (founded in 1968) or Kayson (founded in 1975),
  - Affiliates of the Bonyads (religious foundations) such as Samen Oil, affiliate of Astan e Qods e Razavi, the country's richest religious foundation,
  - Newcomers to the sector after the rise of hardliners, after the election of former President Ahmadinejad. These newcomers include the Revolutionary Guard's engineering arm, Khatam al Anbiaa or Ghorb.[32]

Some of these companies lack the experience in conducting projects on a commercial basis for the energy sector. Moreover, if contracts continue to be awarded, by default or on purpose, to Iranian companies, foreign players would be mostly welcomed only as subcontractors, which is not attractive for potential investors. This situation would not enable rapid growth in capacity because most local contractors are overstretched. In the specific case of Khatam al Anbiaa, even

[32]In 2006, Khatam al Anbiaa was awarded the contracts for the development of phases 15–16 of South Pars without competitive bidding. It was also awarded the strategic IGAT-7 gas line (Assayuleh—gas from South Pars 9–10- to Iranshahr). IGAT-7 is particularly important because it is the cornerstone of NIGC plans to extend gas infrastructure to the border with Pakistan anticipating future gas exports.

those companies accepting to act as subcontractors will have to weigh the opportunity against the political risk of working for the engineering arm of an organization that is explicitly targeted by US sanctions.[33]

In other words, the Israeli and Iranian experiences highlight how difficult it could be to introduce wide reforms in countries with a well-established industrial tradition, even when the three main drivers for reforms identified earlier are relevant. These two countries' experiences also show the limits of reforms where there are multiple of stakeholders. These two points will be detailed later in this book while discussing the attractiveness and the feasibility of the reforms.

3.3 Interconnections and Regional Integration: Sometimes an Enabler, Rarely a Driver

The most plausible way to a [European] common market is the creation of regional markets
D. Finon (2001)

This chapter would not have been complete if it disregarded the various interconnections, existing and planned, between the different countries in the region and their neighbours. Is there a strategic upside in cross-countries' interconnections? Could the Medring, designed to link southern European countries with the Maghreb and the Near East, become the catalyst for the long sought-after common Mediterranean market? Projects are underway in the region, and there are even plans to ultimately link the Medring to the Gulf Cooperation Council (GCC) grid to take advantage of seasonal differences. However, the credibility of these ambitious, cross-border and cross-regional transmission projects is frequently questioned given the high capital cost required for each phase, and the prioritization of indigenous generation projects meant to ensure security of electricity supply. There are three regional interconnection projects, at differing stages: the Medring, the GCC grid and the EJLST grid (Egypt, Jordan, Lebanon, Syria, Turkey). These interconnections are important because:

(a) They would link countries which have introduced different market rules (for new entrants for instance), with different dynamics.
(b) They create markets of different sizes which could allow additional reforms in the future—reforms that would have been irrelevant in smaller markets.

[33]Iran's Revolutionary Guards was not listed on one of the latest Foreign Terrorist Organizations (FTO) list dated April 2008. However, since October 2007, the organization is on the list of "entities of proliferation concerns and support for terrorism concern" under E.O. 13382. At the time of writing the fate of the sanctions regime, including the Joint Comprehensive Plan of Action (JCPOA) adopted in October 2015, was uncertain.

The purpose of this section is to understand the impact these regional interconnections could have on the different electricity reforms being undertaken at the national levels.

The motivation for regional electric interconnections followed the same pattern as the one for different Arab political and economic groupings: it has come and gone.[34] Depending on the various periods of the Arab world's history, each country's energy policy was governed by its closeness to socialism or capitalism, by oil prices and by its own domestic priorities. More importantly, each country's energy policy was different from his neighbours'. However, the experience of the electric interconnections in the MENA region contradicts two intuitive hypotheses:

- One needs to have a solid political grouping with working institutions—such as the European Union, with its Commission, Parliament and Association of regulators—to make a regional electric grid materialize.
- A strong political or economic grouping, with a shared economic vision, could help electric interconnections projects move more quickly.

In general, the pace of development of a regional electric interconnection depends also on the international energy context, and on energy issues which would be specific to the region: shortages, price evolution, fuel mix policies and the debate on nuclear power.

The benefits of electric interconnections for MENA countries started to be obvious to some policy-makers: the transfer of peak load across time zones and the exploitation of the differing load profiles between groupings with different geographies and consumption patterns. For example, the peak load in Eastern Mediterranean countries and Egypt occurs in the winter while the peak load in Gulf countries is in the summer. Ultimately, a connection would allow for reserves margins to be partially shared, and the synergies would enable savings on capital costs need for future generation projects. Moreover, hydrocarbons-importing countries would get access to electricity produced at lower unit cost, particularly during peak periods when dispatch rules would call on the most expensive units. Finally, the temptation to be able one day to be connected to the European market, and maybe to export electricity to it, was also taking shape in some leaders' minds.

The challenges were technical, financial and sometimes cultural. The first needed step was to establish national grids—Saudi Arabia, the UAE, and Oman—to overcome the technical challenges resulting from different operational systems and to synchronize the grids. The second challenge is probably the most important: the cost, of this first step and of the interconnection itself. The last challenge is more cultural in

[34]"Regional cooperation, regionalism and regionalization in the Middle East are usually considered to be weak and rather ceremonial. However, since September 11, 2001, a new regional order is emerging and the impact of geostrategic changes in the international environment has yet to be satisfactorily studied. With older regional organizations suffering from weaknesses, new forms appear to be developing and flourishing, due either to European support or growing sub-regional identities". Harders and Legrenzi (2008) tackle the critical question of why the Middle East—unlike most other regions—has failed to develop robust institutions to promote regional security and development, and explore "the limits and opportunities of cooperation in a troubled region".

nature and might have actually helped the process of reforms in the region—namely, to consider electricity as an internationally tradable commodity dependant on external markets' fundamentals and not as a strategic asset under the sole supervision of the State.

The first real impetus for regional interconnections was given by the Arab Fund for Economic and Social Development (AFESD),[35] when it commissioned in the 1990s three major studies on the feasibility and conditions to build power grids in MENA:

- To link the countries on the eastern side of the Mediterranean: Egypt, Jordan, Syria, Iraq and Turkey
- To link the Maghreb countries (members of the Arab Maghreb Union): Morocco, Mauritania, Algeria, Tunisia and Libya—Algeria, Morocco and Tunisia were already exchanging electricity at the time
- To link all 22 Arab League states.

The last report concluded that the project was feasible. The other main conclusion was that, without the grid (for which the cost was estimated at $6.5 billion), Arab countries would have to invest more than $100 billion between 1995 and 2010 at constant 1995 prices in new generating capacity to cover demand.[36]

In parallel, the project of the GCC grid was being launched under the auspices of the Council (the Gulf Cooperation Council). However, a comparison between the experiences of the Maghreb and the GCC grid projects shows that strong political groupings do not automatically translate into rapid progress on electric interconnections. In reality, these projects appear to be driven by the financing question and by a shared electric vision among the participating parties.

The GCC grid suffered from reduced budgets during a period of low oil prices and from an unclear energy vision in 1990s. The study commissioned by the electricity committee of the GCC secretariat concluded that a grid would reduce the costs requirements for new generation by $3.5 billion up to 2010. At the same time, the GCC acknowledged that there are no substantial gains to be made as peak loads overlap on a seasonal basis, and member states did not have at the time sufficient reserve margin to share. The plan for the GCC grid was divided into three stages: the first linking the northern countries was originally expected to be

[35]The Arab Fund for Economic and Social Development (AFESD) is an autonomous regional Pan-Arab development finance organization. Its membership consists of all States who are members of the League of Arab States. The Agreement establishing the Fund was adopted by the Economic and Social Council of the League of Arab States on 16 May 1968. The General Secretariat of the League of Arab States declared the Agreement effective as of 18 December 1971. The first meeting of the Board of Governors was convened on 6 February 1972, and the Fund commenced operations in early 1974. Its function is to assist the economic and social development of Arab countries through (1) financing development projects, with preference given to overall Arab development and to joint Arab projects; (2) encouraging the investment of private and public funds in Arab projects; and (3) providing technical assistance services for Arab economic and social development.

[36]The analysis was based on a projected annual demand growth rate of 6.6% between 1995 and 2000, and 6% until 2010.

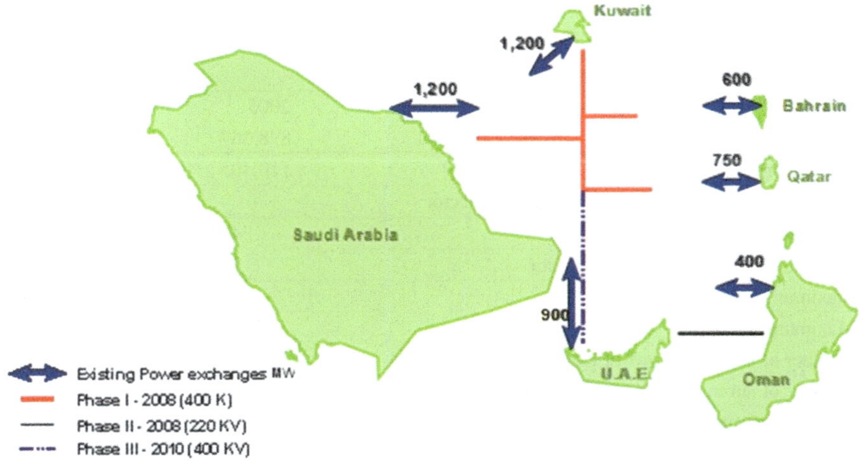

Fig. 3.4 The GCC grid

finished in 1998, the second stage linking the UAE and Oman in 2003, while a third stage would link the two groups. All stages have been delayed. The first phase was only completed in the summer of 2009, 11 years later.

Given the similar features of GCC electricity fundamentals (peak periods, demand volatility), one would question the rationale of the GCC grid. First, the political dimension should not be underestimated. The project did not need to have a solid political affiliation with working institutions (like the European Union, with its Commission, Parliament, and association of regulators) to materialize.[37] But the electric interconnection project has become more important as other pan-GCC economic cooperation projects, with the exception of the Dolphin gas project, were compromised or delayed (shared nuclear reactor, monetary union, railway). Often electric interconnections allow the transfer of peak load across time zones and the exploitation of the differing load profiles among different geographies and consumption patterns. However, that is not the case for the GCC countries. The main purpose of the grid is to help reduce blackouts as power can be provided on an emergency basis.[38] As in other interconnected markets, electricity price convergence is not necessarily profitable to all consumers and all parties. Ultimately the authorities hope for reserves margins to be partially shared, to decrease capital costs needs.[39] Some would get access to electricity produced at a lower per unit cost, particularly during peak periods when dispatch rules would call on the most expensive units (Fig. 3.4).

[37]Treaties were required as the GCC is not a legal entity.

[38]Several hours of emergency power a few times per year for each country, to be paid back in kind, quantity for quantity, with penalties imposed if this does not occur within a certain time frame.

[39]Initial studies in 1991 estimated that 5 GW could be saved by interconnecting the six GCC countries over a period of 15 years.

Table 3.6 Energy exchanges through the Maghreb–Spain interconnection and general features of the 400-kV ac interconnector

Energy exchanges through the interconnection			
Period	2004	2005	2016
Spain to Morocco (MWh)	1,567,641	898,367	4,949,400
Morocco to Spain (MWh)	21,190	110,405	0

General features of the 400-kV ac interconnector	
Rated voltage (ac/dc)	400/450 kV
Maximum transmission capacity (ac)	2 × 700 MW
Maximum overload capacity	2 × 900 MW (for 20 min)
Maximum sea depth	620 m (2016 ft)
Number of submarine power cables	7 (6 + 1 spare)
Number of submarine fiber-optic cables	4
Submarine route length	26 km (16 miles) 29 km (18 miles)
Land section length Spanish side Moroccan side	2.1 km (1.3 miles) 0.25 km (0.16 miles)

Source: GRANADINO, Ramón, MANSOURI, Fatima, *A bridge between two continents,* Transmission and Distribution World, May 1, 2007

Conversely, the progress on the Maghreb grid was more rapid, despite the frailty of the Arab Maghreb Union as a political organization. Algeria and Tunisia were exchanging electricity since the 1950s. Morocco and Algeria started border electricity exchanges in the 1980s. In 1988, the Comité Maghrebin de l'Energie Electrique (COMELEC) was created and has met regularly since then. Until the mid 1990s, Algeria supplied electricity to its two neighbours, mainly to cover emergency load. Then, the civil war of the 1990s resulted in poor maintenance and sabotage which hit the Algerian transmission system and negatively impacted cross-border trade.

Despite Algeria's difficulties, COMELEC created in 1995 a regional power plan and implemented it. It created three commissions to follow technology, planning and human resource issues. Regular talks between the regional utilities resulted in several cooperation programs, mainly in rural electrification. The other interesting feature in the Maghreb grid is Morocco's link to Spain and Europe. The first 400 kV line was commissioned in 1997, with commercial operation starting in 1998, on the basis of short term bilateral agreements: Morocco's ONE had a 5-year purchasing agreement with Red Electrica de Espana (REE) for a guaranteed 90 MW per year, with the option to sell up to 210 MW to commercial operators in Spain. Since then, it has been mainly Morocco who has been importing from Spain.

On the other hand, these electricity flows have contributed to network stability and removed constraints in the Spanish southern area. In 2006, a parallel line was commissioned (Table 3.6).

Fig. 3.5 The Medring: North Africa-East Med-Europe

The Morocco-Spain link played also a role in the reinforcement of the Maghreb grid. These regional interconnections became part of a broader vision, within the more ambitious Mediterranean Electricity Ring (MedRing) project. MedRing is designed to interconnect the electrical networks of all Mediterranean countries: Morocco, Algeria, Tunisia, Libya, Egypt, Jordan Lebanon, Syria and Turkey with Greece, Italy, France and Spain.

The economic feasibility of direct interconnection between Algeria, Tunisia and Libya to Europe (Spain and Italy) was studied on several occasions. Nevertheless, it appeared that it was more economical (five times according to some studies) to import gas from North Africa and generate electricity in Europe than to import electricity. However, most of these conclusions were drawn in the late 1990s, early 2000s. Today, the introduction of CO_2 prices means that the economics merit a fresh look. This issue merits to be a possible topic for future research (Fig. 3.5).

In any case, at the country level, project coordination is decisive when it comes to develop integrated projects. This coordination is *de facto* easier in countries where the central government or a mandated central authority takes on the role of approving, planning and coordinating all major development projects. This coordination is even more crucial when it involves a group of neighbouring countries: power reserves margins in Middle Eastern countries range from 3 to 24%. With such differences, a regional energy integration policy appears to be necessary, even if policy makers could be distracted by internal priorities, including by their own national electricity investment plans.

With a group of state-owned utilities, the nature of the institutional set-up in each country should not act as an impediment for cross-border interconnections. As the different experiences in the region showed, the usual obstacle is funding and lack of common vision. However, reforms in the power sector in a specific country could be create challenges when developing a vision for the region as a whole. Deregulation in a country A poses the question of the "reciprocity clause"—whether the utility of the neighbouring country B could gain ownership or access to generation or transmission assets, and trade in market A while utilities in country A can not enter the market B still governed by a monopoly. Therefore, interconnections could create the environment for "reforms dominos", as reciprocity could push for some limited reform in a previously closed, vertically-integrated system.

Although most countries in the region introduced some competition at the generation level, there has been no example yet of market penetration by one regional utility in the region. Most IPPs in the region were built, owned and operated by the traditional global utilities (IP, AES…). Until recently, the only IPP being operated by a regional player was Jorf Lasfar in Morocco, mentioned earlier. Jorf Lasfar was part of the assets transferred to TAQA in 2007 when it acquired CMS Generation, the independent power production unit of US-based CMS Energy. And Morocco's ONE was the first national utility in the region to display international ambitions, even if the strategy was to target specific projects in Subsaharan Africa. In general, there has been an increasing number of companies, usually from Gulf countries, investing in power assets globally. Examples include:

- **TAQA**[40]: the trajectory of the Abu Dhabi National Energy Company (TAQA) is unique. The company was formed in 2005 to transfer the ownership of the previously state-owned power generation and desalination assets of ADWEA to the private sector (six IWPPs). Its IPO was issued at its creation. It is a Public Joint Stock Company and is 51% state-owned through ADWEA. It is listed on the Abu Dhabi stock exchange. The company's impressive growth is attributable mainly to global acquisitions. The strategic objective of the company is to "*deliver profitable growth in multiple markets where we can access stable cash flows, which complement and enhance our existing diversified energy portfolio*".[41] TAQA's assets include power generation, transmission and distribution assets but also upstream and gas storage facilities.

- **Gulf Investment Corporation**: the GIC has a completely different spirit. It was created in 1983 by the six GCC countries, as a regional financial institution to stimulate private investments and to fund projects that would enhance economic and social development. Throughout the years, the institution played the role of financial advisor to several entities including various utilities in the region. Its very first direct involvement in a power project as a developer occurred in 2004 when it partnered with Tractebel to win Bahrain first private power project, al

[40]"We started as a domestic power company and in about 8 months we went from that to being present in ten countries and having 2000 employees. The whole idea is to take the diverging corporate cultures because we are integrating BP's culture, CMS Generation's culture and Northrup's culture into one company, and we want to develop the TAQA way of doing business. This revolves around operational excellence, performance excellence, environmental excellence and, indeed, employee development excellence." Peter Barker Homek, CEO of TAQA, 2007.

[41]TAQA's CEO Peter Barker Homek, February 2009.

Ezzel. Its performance has been fluctuating over the years. The company's management states that even if GIC's financial goal is to maximize shareholder value through earning competitive rates of return, its underlying "principle" is to support projects and economic activity.

- **Mubadala**[42]: Another Public Joint Stock Company headquartered in Abu Dhabi – 100% owned by Abu Dhabi. Since its creation in 2002, its main commercial strategy has been to build a portfolio of long-term capital-intensive investments that deliver strong financial returns. Mubadala is mainly known for the development for the first cross-border natural gas network in the region, the Dolphin Gas project. But it has invested in various sectors and industries in different parts of the world. Therefore, it is mostly an investment vehicle of the government but with a commercial and economic orientation.

- **Oman Oil Company**: OOC is the investment vehicle of the government, a commercial state-owned company pursuing investments, primarily in the energy sector, in the country and internationally. It was created in 1996 and its main objectives are related to the diversification of the Omani economy, job creation and revenue generation for the government.

However, most are not utilities and had yet to gain the technical and managerial experience to be able to compete in international IPP tenders. They might be financially robust, but they are often technically questioned. On the other hand, the flurry of projects in the region created some local developers which gained advantage from focusing on their home country. Saudi Arabia is a good example in this regard. The portfolio of Saudi power projects is large enough to provide opportunities to a wide array of players. Local companies like ACWA power, National Power Company or Sadaf were able to compete and win IWPP projects.[43] Newcomers might assert themselves in the race for power projects if they partner with the right international players. Finally, national utilities, who might consider "going international", usually struggle to maintain a positive balance sheet given the prevailing subsidized tariffs. They depend on their governments to have their development plans and investments approved, and are frequently in competition with other sectors (education, health, defence…) in national budgets.

In conclusion, there are three main lessons to learn on the impact of interconnections on electric reforms:

- Established political groupings do not automatically act as a catalyst for interconnection projects, not to mention electric reforms.

[42]Abu Dhabi's third investment vehicle in energy assets is International Petroleum Investment Co (IPIC). The three companies to not compete with each other.

[43]Acwa won its first IWPP at Shuaiba in 2002. Sadaf executed the country's first ever IPP in 2004 in Jubail.

Table 3.7 The state of reforms in the region in 2008

	Relative openness to IPPs	Effective move towards deregulation	Single Buyer Model	State monopoly (or <2% of private generation)
Morocco	Open		X	
Algeria	Open		X	
Tunisia	Open		X	
Libya	Closed			X
Egypt	Open	X		
Israel/ Palestine	Open			X
Jordan	Open		X	
Lebanon	Semi-open	X		
Syria	Closed			X
Saudi Arabia	Open	X		
Yemen	Open		X	
Oman	Open		X	
UAE	Open		X	
Qatar	Open		X	
Bahrain	Open		X	
Kuwait	Closed			X
Iran	Semi-open			X
Iraq	Semi-open			X

- Regional cooperation between national utilities could make progress easier. However, that would be challenged when scattered deregulation movements start.
- However, interconnections could create the environment for "reciprocity clauses" to be negotiated between neighbours, and push some limited reform in a previously closed market.

3.4 Then What for the Future?

In 2008, reforms slowed down and the few moves towards more deregulation were generally reversed (see Table 3.7). This deceleration coincided with a new wave of "resource nationalism" witnessed in oil-resource holders during the same period. Examples in the region[44] included the reversal of the hydrocarbons law in Algeria or

[44]Other examples include Venezuela, Russia and Ecuador. For more background on resource nationalism, refer to Johnson (2007) and Stark (2007), for instance.

the tightening of the EPSA terms in Libya.[45] The deceleration of deregulation occurred in non-oil producing countries as well. In reality, it mirrored the "reregulation" process that occurred in some already-deregulated markets in other parts of the world. As of 2016, the international re-regulation trend, the global financial crisis, the Arab Spring all contributed to slow down or reverse the few emerging movements of deregulation beyond the Single Buyer Model (see Table 3.8 bis).

It has been generally admitted that the movement toward less regulation and more reliance on market processes in the electricity industry has enormous potential benefits. A series of crises including the California crisis reminded the industry that deregulation also bears potential risks. Furthermore, transmission can be a real bottleneck as levels of transmission capacity that may have been adequate under regulation might not be able to withstand full competition. In some areas, restructured electricity markets proved more costly, in the short term at least, for consumers than the regulated predecessors. The difficulty of measuring the long-term benefits of improved investment decisions discourages potential candidates to deregulate further.

In the region, wholesale competition started to show its limits when applied in an environment of heavily subsidized tariffs (Oman, the UAE). The transfer mechanism to the final consumer of production costs' reduction in case of efficiency gains or of wholesale cost increase does not exist or does not occur effectively. The production cost fluctuation has only an impact on the government's budget; the "triangle of opportunism"—government, companies, consumers—turns into a simple dual relation between a government and (a few) companies. In other words, the final consumer does not see the gains registered by the introduction of wholesale competition. On the other hand, Morocco displayed the limits of a weak or absent regulation in the framework of a liberalization process. In general, the absence or the weakness of an independent regulator who would monitor and regulate the whole process can jeopardize further deregulation. The core issue is that price-responsive

[45]In Algeria, the 2005 hydrocarbons law was hailed at the time as a major step toward full liberalization of the hydrocarbons sector. The law clarified the state's role in the oil and gas sector as policymaker and regulator through a set of new, independent agencies. It also sought to allow free competition through the elimination of monopolies. It established the framework of a liberalized operating environment that aims to guarantee fair and equitable treatment of all operators, whether public or private, domestic or foreign. In the 2005 version, Sonatrach's "right" to a 50% equity stake in any project is replaced by an "option" to take 20–30% interest in case of discovery. In July 2006, a presidential decision partly reversed the law's liberalization trend by instituting a windfall profits tax and by giving back to Sonatrach a mandatory 51% interest in any future E&P contracts as well as in pipeline projects. In Libya, the EPSA bid round of 2006 offers a contrasting approach, in comparison to the unilateral approach of other countries. In accordance with the standard Exploration and Production Sharing Agreement, Libya invited bidders to stipulate the share of production they would accept. The result was offers in the range from 7.8 to 13.5% with one offer accepted at 18% but reduced in subsequent negotiations to 12%. This is very low compared with previous rounds and contracts. Other criteria on which bids were evaluated included parallel investment proposals (particularly social or training investments) and additional advantages such as the increase of local content.

Table 3.8 Bis: the state of reforms in the region in 2016

	Relative openness to IPPs	Effective move towards deregulation	Single Buyer Model	State monopoly (or <2% of private generation)
Morocco	Open		X	
Algeria	Open		X	
Tunisia	Open		X	
Libya	Closed			X
Egypt	Open		X	
Israel/ Palestine	Open			X
Jordan	Open		X	
Lebanon	Semi-open			X
Syria	Closed			X
Saudi Arabia	Open	X		
Yemen	Open		X	
Oman	Open		X	
UAE	Open		X	
Qatar	Open		X	
Bahrain	Open		X	
Kuwait	Semi-open		X	
Iran	Semi-open			X
Iraq	Semi-open			X

demand is more critical in less-regulated markets. In the absence of a link between price and demand, one cannot even rely on price to regulate the market.

The crisis of confidence that hit the economy in the wake of the 2008 financial crisis and the wide questioning of the value of the role of the market comforted the bearish attitude towards further liberalization in each country of the region. However, at the time of writing, several countries were reforming their retail electricity tariffs, with the view of making all transactions within the sector economic. Also, Saudi Arabia, the only exception, is attempting to progress its long overdue utility reform, driven by the National Transformation Plan of Vision 2030. As of 2018, a national transmission company and a principal buyer were created and operational, but progress on the remaining parts of the Saudi utilities restructuring plan has stalled.

MENA countries are not isolated, and it appeared important to analyse what impact (if any) the regional interconnections could have on the different electricity reforms being undertaken at the national levels. The experience of MENA electric interconnections is all the more interesting that it contradicts two intuitive hypotheses: that one needs to have a solid political grouping with working institutions to make a regional electric grid materialize, and that a strong political or economic grouping, with a shared economic vision, could help electric interconnections projects move more quickly. The analysis confirmed that established regional political groupings do not act as a catalyst for interconnection projects or for mention electric reforms. Actually, it is the regional cooperation between national (state-owned) utilities which could make progress easier, something that would be challenged when scattered deregulation movements start. However, interconnections could create the environment for "reciprocity clauses" to be negotiated between neighbours, and push some limited reform in a previously closed market, even the number of companies which can really compete on a regional basis is limited.

In a nutshell, the success of the wholesale competition model, its simplicity and its relatively painless implementation contributed to its adoption across the region. All the constraints detailed above indicate that most countries would not be ready to prepare for the implementation of retail competition, or further deregulation, in the absence of clear benefits, and assuming retail competition is suitable.

3.5 Summary

Electricity-sector reforms started in the mid-1990s in the region, following the "second" global wave of electric reforms which occurred in Latin America and Eastern Europe. Oman and Morocco led the way in the early 1990s, and they were soon followed by Abu Dhabi and the majority of other MENA countries. As in other regions around the world, several factors played a major role in urging the governments to restructure their electricity sectors: growth in electricity demand, the sense of urgency and the inability of the governments to respond efficiently, to cite a few. It is conventionally admitted that, in the region, oil prices are an important factor influencing economic reforms. This was probably true during the last two decades of the last century but not afterwards, particularly as the world wavered between "too little" and "too much" regulation. Even countries which were tempted by quick privatization revenues understood that privatizing assets without first investing effort in restructuring might not be as lucrative as originally thought. The need to build consensus among decision-makers guided electricity reforms, particularly in gulf countries. But this consensus was sometimes led by financial and industrial interests of powerful individuals. In other countries in the region, where interests are concentrated but not mutualised (North Africa), a general suspicion towards anything related to the introduction of the private sector would hamper the process of reform or privatization. The productivity of most vertically integrated monopolies—or quasi-monopolies—in the region is comparable to that of a western vertically-integrated monopoly. However, the limited financial capability of some could be problematic. It hampers their capability to raise funds, creating performance problems when the utility is not backed by a financially strong State.

Holburn and Spiller (2002, p. 465), argue that *"having institutions right"* is more important than *"having the structure "right"*.

Chapter 4
Evaluation Step 1: Impact on Production Costs (Performance) and Investments (Incentive)

It is clear that deregulation is a high risk choice. Those jurisdictions that have not yet deregulated electricity generation need to think long and hard before they go ahead. Those that have done so need to figure out how to minimize the downside potential of the journey on which they have embarked.
Newbery (2002).
En périodes de crise, le fait d'avoir des acteurs (transport, distribution, commercialisation, production) complètement séparés rend la coordination de la gestion des crises très difficile ("During crises, having completely unbundled actors (transmission, distribution, retail, and generation) makes the coordination of the management of crises very difficult". Interview des Echos, 3 October 2003).
François Roussely, former President, Electricite de France (EDF).

Abstract Liberalization does not necessarily reduce electricity prices. In the case of directly subsidized systems, the only possible "price" indicator would be the wholesale cost. In countries which introduced competition in generation and implemented the single buyer model, wholesale tariffs or Bulk Supply Tariffs could be the best proxies to measure possible efficiency gains in the sector. The evaluation will also depend on the chosen method of regulation ("rate of return" or "RPI-X"). The second step of this evaluation consists in analysing whether reforms contributed to attracting private investments. The MENA region was able to attract a portion of global private investments, although not as much as other emerging regions (e.g. Latin America). The fuel cost advantage made the region relatively attractive for power developers, but it was not sufficient and not sustainable.

Keywords Regulation · Private investments · Attractiveness · Incentives

L. Benali, *Electricity-sector Reforms in the MENA Region*, Perspectives on Development in the Middle East and North Africa (MENA) Region,
https://doi.org/10.1007/978-3-319-96268-9_4

The starting point of this analysis lies in an observation: electric systems in the region do not automatically fulfil the favourable conditions to deregulation presented by Joskow (2000): (a) demand stagnation and chronic overcapacities, as the sector gets mature; (b) an advanced stage in production, transport and information flow management techniques' evolution. However, a series of factors lie behind reform movements: poor performance of the state-owned/run electricity sector (costs, access to service, reliability), inability of the sector to self-finance capital, or even O&M expenditures, the need for government to capture revenue through the change of tariff structure/decrease or removal of subsidies and/or through privatization of assets. It is generally perceived that any monopoly structure has one main weakness: it does barely offer mechanisms of cost reduction, because of the lack of efficiency incentives.

Once the need to reform the sector identified, it is important to decide when to proceed, how much is needed, and who will finance the response to these needs. Even when the latter is ultimately externalized, the planning function remains critical. In completely deregulated markets, price is at the centre stage as it is supposed to do most of the economic work: dispatch and allocate resources, chase market power, send the signal at the right moment for new supply and reduce consumption when shortages are threatening. Whether it is actually able to perform its "economic duties" is another debate. In less deregulated markets, the transition from a monopoly structure to any other structure often raises the same questions about the benefit of the reform, particularly since it usually comes at a cost (institutional redesign, political, contractual costs), but it is generally difficult to assess that benefit.

Joskow (2006) and Littlechild (2006) offer the ten Textbook regulatory elements for a successful competitive reform of electric industry:

1. **The privatization** of the electric industry. It is understood at the same time as a means to introduce competition, to give strong incentive of the industrial and financial performances and to work as a barrier to the governmental economic and political pressures.
2. **The vertical separation** of the competing segments (production and supply) of the segments having to remain under public regulation (transmission grids and distribution networks). This separation can be carried out structurally or legally and must guarantee that cross subsidization between the two segments will be impossible and that the access to the grids will not be discriminatory.
3. **The horizontal dismantling** of the competing segments should create sufficient number of participants in the various markets if any (wholesale and retail). This is necessary to guarantee *ex-ante* reasonable competing mechanisms.
4. **The creation of an independent management of the transmission Grid**. Its functions include: real-time balancing of electricity; guarantee of quality (frequency voltage and stability). This separation withdraws the temptation of discriminatory practices in the operational management of network.
5. **The creation of electricity markets in real times** coupled with competing mechanisms of allowance of the physical constraints of electric transport are is

also necessary. They allow that the risks of production and consumption are managed effectively.

6. **Rules of the game are needed for setting**: (a) Third Party Access on nondiscriminatory basis; (b) effective allowance of the transmission capacities; (c) signals of localization and connection for the new entrants.
7. **The identification and the separation of the costs of the electric supply** must be made: thus each consumer has to know in its consumption: (a) the cost of the networks, (b) the cost of the supply of energy and finally (c) le cost of the services associated with the electric supply.
8. **The creation of an independent regulator** is a crucial point of the reforms: regulator must have access to sufficient information (costs, quality of service, compliance with the rules of competition) and hold the sufficient authority and means to effectively implement the monitoring of the operators on the networks activities and in competitive activities.
9. **Mechanisms of transitions** and ex post amendments of the initial scheme of the reforms are always necessary.
10. **Coherence of decisions in time**. The absence of decision-making in the first nine topics is largely prejudicial. Electric reforms are a multi-year evolutionary process of improvements and correction of errors, and potential omissions.

The first step in any reform evaluation is to analyze the effect on price for end-users. Several studies [Nagayama (2009) for example] concluded that "higher electricity prices are one of the driving forces for governments to adopt liberalization models", mainly in order to increase the performance of the sector. However, they also concluded that the development of liberalization models does not necessarily reduce electricity prices. In fact, contrary to expectations, empirical analysis in various markets shows that prices tend to rise. The results of the reforms have differed greatly from country to country, even when the scope, the policy orientation, and the objectives were similar. The transactional complexity of the electric industry is common in all countries, while the institutional frameworks are very country-specific.

In the case of directly subsidized systems, the only possible "price" indicator would be the wholesale cost. Indeed, when end-users are subject to flat, below-cost prices, the performance of the system post-reform could be measured by the evolution of costs at the wholesale level. This would enable the measurement of the hypothetical cost reductions if any, except possibly in distribution companies when these receive the entire subsidy.[1] One difference with systems where costs are entirely passed through to the end-user would be that these reductions would benefit the State—in subsidized systems—instead of directly benefitting the end-user. Even in heavily subsidized systems, cost reductions would benefit citizens since the reduction in the volume of subsidies would enable governments to invest in other areas. However, price-insulated end-users would have limited impact in the sector,

[1]World Bank proposes other means that could serve to measure the performance of distribution companies.

as they would be denied the status of institutional stakeholders. They would therefore be unable to influence the making and the future of the reform, at least in an institutionalized way. However, in countries with active parliaments, citizens could influence the reform program of the government, through the review of the national budget and the size of subsidies or the questioning of ministers.[2]

The second step in the evaluation exercise will be to look at the impact on investments. i.e. whether the reform program enabled investments, and whether it encouraged investments at the right time, at the right place, with the right technology, and hopefully by the right players. In theory, in deregulated markets, the price is supposed to do all the economic work, including sending to investors the signal to invest in additional generation capacity. In reality, empirical experiences show that other factors get in the way and distort this desired theoretical outcome. The Brazilian example (see textbox) demonstrates the need to define all the rules that will govern the new structure of the market. It shows that investments can be dramatically affected by an unclear, hesitating and incomplete reform process. It provides evidence for Joskow's argument on the role of generation competition and auction markets: "the performance of these auction markets depends critically both on there being robust competition among generation suppliers and the implementation of a set of auction rules that are compatible with the physical operating constraints on the system and do not facilitate gaming and market power problems that may be exacerbated by these physical constraints".

The Electric Reform in Brazil and the Impact on Investments

In Brazil, the reform process has never been completely implemented, partly because of changes in the macro-economic context (economic instability, exchange rate devaluation in 1999). The government was unable to reach an agreement on the appropriate industry structure in the generation, dominated by five companies controlled by the same holding company, Eletrobras. The difficulty in defining the industry structure and an adequate regulatory framework taking into account the peculiarity of the Brazilian electricity industry affected other dimensions of the reform, such as the definition of clear rules for the wholesale electricity market. In the absence of such rules, it was not possible to anticipate future price levels. Most of the electricity generated by the hydroelectric companies was sold on a historic cost basis through the initial contracts. These costs were very low compared to the current marginal costs. Therefore, the regulatory framework was not sufficiently developed to forecast future electricity price levels. The amount of electricity negotiated through the wholesale market, furthermore, has not been significant.

(continued)

[2]In MENA, most countries have a parliament or a similar form of citizens' representations. Saudi Arabia and Qatar have unelected "Shoura Council".

The rules to allow the coexistence of hydro (90% of electricity generated, with much lower short-term marginal costs) and new thermal power generation, particularly how they will compete, were undefined. Thermal generation plants cannot beat the SRMC of hydro plants during periods of adequate rainfall. Therefore, it appears important to define rules that allow for new thermal generation and for the optimization of the use of hydropower. Finally, gas pricing policy has been a major uncertainty, with high cost and volatility being real obstacles for the development of thermal gas plants. Imported gas was 40% more expensive than domestic gas.

The peculiarities of the Brazilian electricity system might have made the reform exercise particularly difficult, and the authorities found it hard to define the most appropriate model of reform to encourage investments. However, the halt in the privatization of the generation segment represented an additional disincentive for investments in capacity expansion. The government suspended its own investment program, expecting the companies to be privatized, while private players were not interested in investing in greenfield projects until the privatization process was completed. Investments remained at very low rates as compared to historic levels.

Total investments in the electric sector decreased from an average of US$ 10 billion (constant) in 1975–1985 to only US$ 5 billion in 1995. Investments in power generation fell from US$ 3.1 billion in 1996 to only US$ 1.8 billion in 1999. The shortage of investments resulted in an increasing gap between demand and installed capacity. Between 1983 and 2000, while consumption expanded by 230%, installed capacity increased only by 180%, deteriorating the reserve margin. Investment was also lacking in the transmission segment. Therefore, even though the Brazilian electricity transmission system is one of the largest in the world, it is relatively fragile, with a significant amount of electricity generated very far from the load centers.

This experience and others highlight the limits of the so-called "Washington consensus" for economies in transition.[3] The implementation of the liberalization/privatization process often faces problems which were raised by several critics, including Stiglitz[4] (1998, 1999). First, the consensus' main measures are built on a limited understanding of the institutional change. Adoption of market institutions without a clear understanding of the minimum transaction costs and of the implementation challenges of institutional change can be perilous. Clearly, the functioning of these adopted, sometimes imported, market institutions requires an ad-hoc institutional framework. More importantly, the pre-existing environment of formal AND

[3]John Williamson used the phrase in 1990 "to refer to the lowest common denominator of policy advice being addressed by the Washington-based institutions to Latin American countries as of 1989."

[4]Former vice president and Chief Economist of the World Bank.

informal institutions will need to be taken into account. Second, the crucial question of the sequencing of reforms is possibly ignored: macroeconomic stabilization and inflation control take precedence, while sectorial reforms—mainly to achieve quick productivity enhancement—become secondary and partial. In these cases, massive privatizations appear.

There are cases where more fundamental measures (tariff liberalization, competition introduction) are circumvented or simply ignored in the reform, because they would lead to inflation, destabilize a pre-existing institutional order or be too "painful" to implement. Efficiency incentives fail to be implemented, while privatized monopolies usually retain an important influencing power to limit future regulatory changes. This last configuration is not necessarily a problem as long as it is a known outcome of the reform process, with accompanying corrective measures, and not a result of an underestimation of the difficulties of the process. However, it is assumed that the primary goal of the reform is to enhance productivity (decrease production costs) and enhance the reliability of the system (incentivize appropriate investments).

The first evaluation framework (costs/tariffs and investments) is commonly used as a rule of thumb to assess the impact of reforms through the effect on costs and on investments. In the next chapters, this approach will be complemented with a criteria grid built around four qualitative and quantitative indicators.

4.1 Does the Wholesale Tariff Send the Right Signals?

In more liberalized markets, eligible consumers would bear a price which would be freely established by wholesale markets. For large industrial consumers for example, this could be a departure from the earlier prices negotiated in their long-term agreements with the utility. The evolution of spot prices might be a measure of the sector performance. However, Vassilopoulos and Rapin for example explore some of the behaviours of electricity spot prices including their volatility and their risks which would be transferred from the producers to the consumers. They examine the adverse effects of prices' contamination by derivatives and conclude by advocating bilateral contracts between producers and consumers as a way to manage that risk.

In countries which introduced competition in generation and implemented the single buyer model (see Appendix B), wholesale tariffs or Bulk Supply Tariffs could be the best proxies to measure efficiency gains in the sector. This assumes that all transactions in the sector (possibly except sales to final customers) are on economic-cost basis. The Bulk Supply Tariff is the tariff at which the Single Buyer sells electricity to the Distribution Companies.

Then, countries would decide how to regulate their Transmission and distribution companies. There are two main methods: the "rate of return" method or the "RPI-X" method. Oman and Abu Dhabi apply the RPI-X method, while Jordan uses a Rate of Return method.[5]

"Rate of Return" regulation, a relatively old method, is based on rate of return calculation. It determines the tariffs to be charged by the regulated entity (the Distribution companies for example) so that it earns sufficient revenues that cover all legitimate operating and capital costs, associated with providing its core activities within performance standards, while providing the regulated entity with a fair rate of return on its capital employed.

More concretely, every year, the regulated entity estimates its costs and the regulator determines the rate of return that it deems "fair and reasonable" and establishes tariffs. The latter will be a function of the revenue requirements which in turn depend on the estimated costs and the chosen rate of return:

$$\text{Revenue Requirements} = \text{costs} + (\text{rate base}) \times (\text{rate of return})$$

The regulated entity should be able to recover all reasonable costs of providing services to customers and should not be financially responsible for factors beyond their control. Tariffs, therefore, can be adjusted between the tariff review periods for factors beyond the control of the entity. Tariffs will be set for a specified period of time to encourage companies to improve efficiency and reduce costs and to provide price certainty to consumers.

There are models where the regulated entity faces financial penalties for failing to meet performance standards, in the form of compensation subtracted from the Revenue Requirements. That should reduce tariffs to consumers.

For example, in Jordan, the Electricity Regulatory Commission (ERC) determines the maximum Revenue Requirement for each Distribution company as follow[6]:

$$\sum_{t=1}^{p} RR_t = \sum_{t=1}^{p} C_t + \sum_{t=1}^{p} (RAB_t \cdot RoR_t) - Pen_{p-1} + SF_{p-1} + Adj_{p-2}$$

t the calendar year, p the number of years in the Tariff Review Period, p-1 and p-2 the previous Tariff Review Periods.

RRt = Revenue Requirement for the year t.

[5]During the 1940–1950s, marginal cost pricing became the main rule to regulate public utilities and widely advocated by theoretical economists. It took time to notice that the main caveat of the method was the assumption that enterprises were maximizing productive efficiency, and empirical experience supported this view. Then economists started to look for alternative methods that would change the relevant incentives. The privatization and deregulation waves of the 1980s and 1990s provided the impetus for efficiency and innovation and incentive price cap regulation was established in areas where competition was not yet feasible.

[6]For more details on Jordan's electricity tariff setting, see "*Determining of Electricity Tariff in Jordan*", Meqdad QADOUS, Director of Studies and Economic Researches Department, ERC Jordan, Tariff/Pricing Committee Meeting, February 4–5/2008, Budva, Montenegro.

Ct = all costs, i approved by the ERC for the year t.[7]
RABt = Regulatory Asset Base for the Core Activities for the year t.
RoRt = Permitted rate of return on the Regulatory Asset Base before tax for the year t.
Penp-1 = financial penalties on the entity for failing to meet performance standards in previous tariff review period.
SFp-1 = Shortfall/excess of the entity's estimated costs and returns in the previous Tariff Review Period compared to the figures used in estimating the Revenue Requirement.
Adjp-2 = a final adjustment to figures for last Tariff Review Period.

Then, prior to 2011, the Rate of Return before tax for each distribution licensee will be: 7.5–16% on paid up capital for JEPCO and 10% on RAB for EDCO and IDECO.

After 2011, ERC expects to use the WACC to determine the rate of return for Distribution licensee:

$$\text{WACC} = \text{We} \times \text{Re}/(1 - \text{T}) + \text{Wd} \times \text{Kd}$$

Where:

Re = Cost of Equity and Kd = Cost of Debt
We = Percentage of equity in overall capital structure and Wd = Percentage of debt in overall capital structure
T = Tax rate applicable to the Distribution Licensee's core activities at the time of the calculation.

The rate of return method has been criticized mainly because it does not provide incentives to the regulated entity, or the monopoly, to reduce costs. Instead, it creates a perverse effect known as the effect of Averch-Johnson. In 1962, Averch and Johnson[8] demonstrated that a cost-plus regulated monopoly would be more inclined to overinvest. In other words, as investments increase the value of the asset base, the regulated entity risks pursuing economically unjustified investments that would bias costs estimates and result in higher regulated tariffs it would benefit from. Therefore, Littlechild and Beesley presented in 1989 the British alternative: the price cap method.

The RPI-X was recognized as offering the best combination, in the long term, in that it set prices to be constantly reviewed downward and it encourages a 'good' quality of service to consumers. It was preferred to the "rate of return" method in some countries. These two methods reflect two differing philosophies of regulation

[7]Costs include depreciation and Bulk Supply Tariffs but excluding interest, loan arrangement fees, and losses or profits as a result of movements in the value of foreign loans arising from changes in the value of the Jordanian dinar.

[8]Averch H, et Johnson LL, Behavior of the firm under regulatory constraint. Am Econ Rev 52 (5):1053–1069, December 1962.

as they are associated to different histories. The price cap method establishes a ceiling under which the regulated entity is free to set up its tariffs. The initial ceiling is calculated based on costs, investment plans and demand forecasts. An evolution formula (usually linked to the Retail Price Index RPI) determines annually the price ceiling over the upcoming period (generally 3–5 years). If the entity is able to "beat the cap"—to optimize its costs so that they remain lower to prices, it is allowed to retain its profits. The X factor is the factor of productivity expected from the entity, either based on its own costs or on a benchmark entity (yardstick style).[9]

Supposing that the revenue function R is a linear function, therefore for a given year t:

$$R_{t+1} = R_t + R_t \times (RPI - X)$$

With R_1 defined as the initial ceiling for year 1, the general revenue formula for a year t during the period of control ($1 \le t \le y$) would be:

$$R_t = R_1 \times (1 + RPI - X)^{t-1}$$

The price cap method would incentivize efficiency and cost optimization, and would allow more price flexibility for multi-product entities, like in some MENA countries—cogeneration of electricity and water. But its main advantage is the relatively long period of control which saves costs for the regulator as well. In fact, its benefits are mainly highlighted in opposition to the cost plus method.

In Abu Dhabi, the RPI-X method is used for all the companies which are subject to price controls by the RSB (four companies), except one. A major part of ADWEC's costs consists of procurement costs relating to Power and Water Purchase Agreements and to the Fuel Supply Agreements. The RSB deemed hazardous to link these costs to demand movements and price inflation. The RPI-X regulation of ADWEC is based on the direct costs over which it has some control. The revenues of TRANSCO, AADC and ADDC are determined each year as a function of three revenue drivers. In the case of transmission (TRANSCO), revenue drivers are: a fixed term (A_{pt}, A_{wt}), an amount function of peak demand recorded by the transmission system (B_{pt}, B_{wt}) and a quantity function if the total capacity of the transmission system (C_{pt}, C_{wt}). The value of these revenue drivers varies between electricity and water, but the general principle is the same. In the case of the two distribution companies, the number of clients is used instead of the peak demand, in order to incentivize the distribution companies to push for introduction of meters. This is all the more relevant in the case of water as water tariffs are set at

(continued)

[9]Competition by statistical comparison.

10 AED/kIG, while residences which are not equipped with water meters pay a forfeit of 50 AED per month. The daily water consumption per capita in Abu Dhabi is about 113.3 IG (500 L).
For the three Transmission and distribution companies, RPI is based on the US CPI (Consumer Price Inflation) and the Emirati CPI, as a major part of their expenses consists of imported goods.

Wholesale tariffs are rarely publicly available in MENA. The Regulation and Supervision Bureau of Abu Dhabi has been one of the rare authorities in the region to regularly publish the sector's economics. And even in this case, methods' changes in the calculation make the evaluation exercise difficult. The approval of the Bulk Supply Tariff has been transparent, as the Regulation and Supervision Bureau must agree on the BST at which the Abu Dhabi Water and Electricity Company ADWEC sells electricity to the two distribution companies in the Emirate (AADC and ADDC). All sector costs are passed through to the Distribution companies. The Transmission Company (TRANSCO) charges for transmitting electricity by the Transmission Use of System (TUoS) charges. The Distribution companies incur Distribution Use of System (DuoS) costs and sales (supply) costs. Figure 4.1 shows the various component of the unit cost at the retail level. It is the mandate of the Regulation and Supervision Bureau to review and approve all costs and charges.

The problem with economic regulation of distribution is that it does not consider, or not adequately, the "quality" of the electricity as a measure to rate.

The BST has two components: an energy charge that reflects the SRMC of providing units at different times of day and in different months (the System Marginal Price—SMP), and a demand charge based on a distribution company's

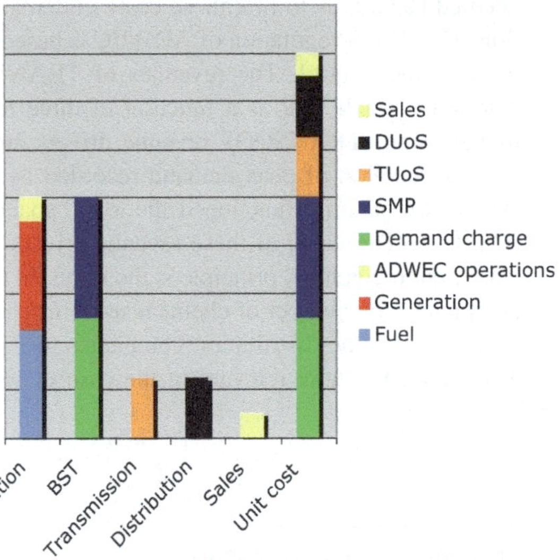

Fig. 4.1 Costs and Tariffs Components in Abu Dhabi. Source: Regulation and Supervision Bureau, Abu Dhabi

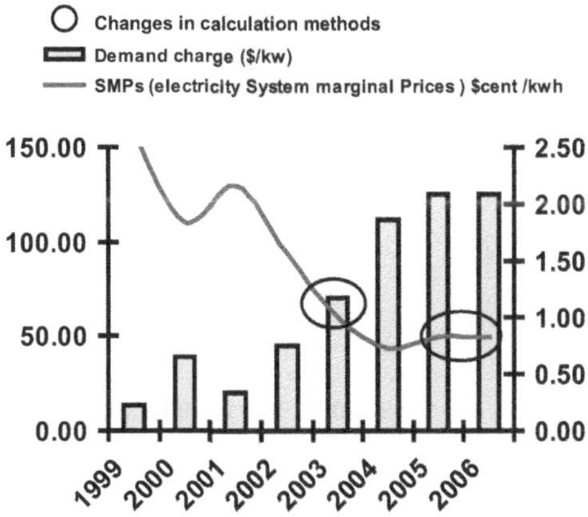

Fig. 4.2 Abu Dhabi's BST components. Source: Regulation and Supervision of the Bureau. Converted to nominal Dollars (1 AED = $ 0.27). Annual reports 1999–2006

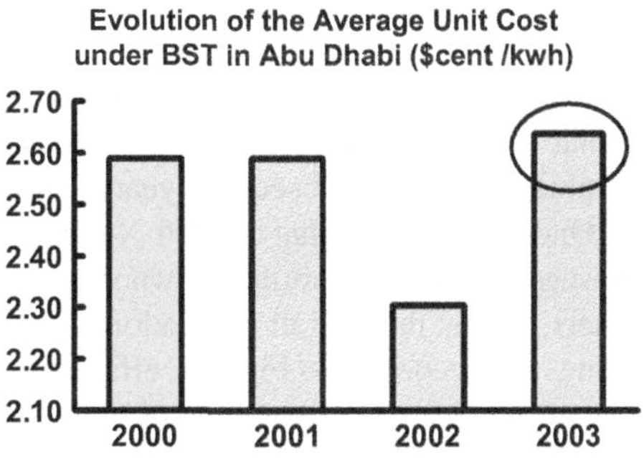

Fig. 4.3 Evolution of the average unit cost under BST in Abu Dhabi ($cent per kWh). Source: Regulation and Supervision Bureau annual reports (2000–2004). In nominal Dollars (1 AED = $ 0.27). No BST data available after 2003

share of demand at the time of system peak demand. At first sight, Abu Dhabi's average System Marginal Price, in $cents per kWh showed initially encouraging signs of decrease (see Fig. 4.2).

However, it is worth noting that a method revision was introduced in 2003 and it is unclear what impact (if any) that revision had on the decrease noticed after 2003, and whether the decrease was structural.[10] As a consequence, it is difficult to draw a definitive conclusion on the BST's evolution before 2003 (see Fig. 4.3).

After 2003, it appears that the BST peaked in 2004 and 2005 ($ 1.1 cents per KWh and $ 1.25 cents per KWh respectively) before decreasing, according to data released from time to time by the RSB. The decrease would mainly be attributed to the diminishing surplus capacity. It is unclear whether the latter could be attributed to slower investment rate or rapid demand growth, or a conjunction of the two.

Therefore, even in the case of Abu Dhabi, it is difficult to assess the actual impact of reform on efficiency, based on publicly available data. The RSB states that the increase in unit production cost was attributable to surplus capacity (estimated at

[10]In 2003, the structure of electricity BST charges was reviewed based on the improvement of the cost reflectivity of the SMP charges and the increase of demand charges.

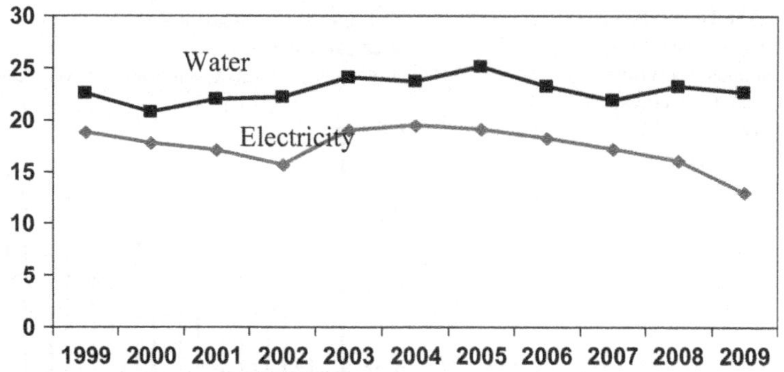

Fig. 4.4 Average unit price trends in Abu Dhabi. Stripped of inflation, based on 1999 prices, all to scale. 2008 and 2009 forecasts. Source: A brief overview of Abu Dhabi's Electricity Sector, a presentation by Nicholas Carter, Director General—Regulation Bureau, PowerGen Middle East 2007

1100 MW in 2004, in addition to the required reserve margin). This at least sent the signal of overinvestment to the authorities. As stated earlier, whether diminishing surplus capacity in subsequent years was accidental or not is unclear.

One conclusion is that the RPI-X regulation is acting as an efficiency incentive for most generators. Government-owned assets will be gradually in competition with the others, so that the overall generation cost would decrease substantially. For the time being, and according to Fig. 4.4, efficiency gains are not important enough to make a real difference. The example of Abu Dhabi shows that current unit costs are comparable to those of 1999.

Oman is the other country that publishes the Bulk Supply Tariffs of its electric system, in addition the economic costs and the underlying subsidies. In this case as well, one cannot conclude that electric reform led to a substantial decrease in costs (Fig. 4.5).

(Bz/kWh)	2009	2010	2015	2016	2017
Economic cost	24.5	24	29.1	30	29.2
subsidy	8.6	7.9	13.5	14.5	11.7

Source: Oman PWP, AER

Indeed, gas prices, which have been maintained at $ 1.5 per MMbtu for power generators are being reviewed. Moreover, the impact of capital costs inflation during the 2003–2013 period was passed through as new plants came online.[11] In the 1990s,

[11]CERA's capital cost index shows that power capital costs nearly doubled between 2000 (100) and 2007 (194). It even exceeded 230 during the summer of 2007. It reaches 171 in 2007 if nuclear is excluded.

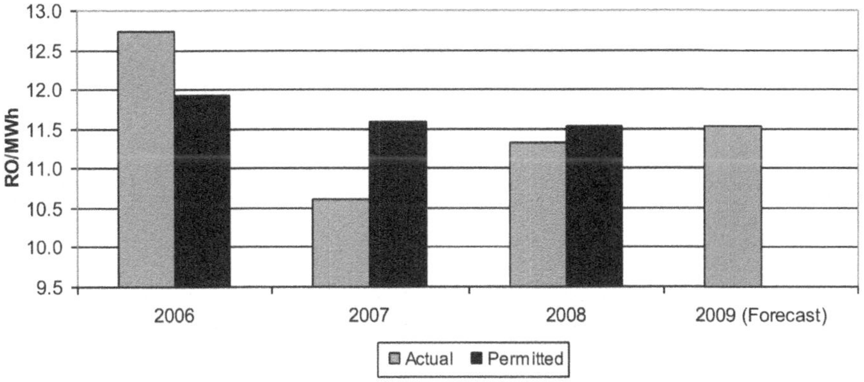

Fig. 4.5 Average BST calculated from actual & permitted revenue 2005–2009, Economic cost and subsidy post-2009. Permitted Revenue reflects actual allowed costs for year (ex post); variations between Permitted Revenue and Actual Revenue are adjusted in next year. Therefore, Permitted Revenue provides best indication of underlying trend. After 2009, AER started publishing on an annual basis the economic cost of the system (below the MIS) and the subsidy

Table 4.1 Examples of electricity and water prices in projects' bids

	SHUAIBA (2005)	MARAFIQ JUBAIL	SHUQAIQ	RAS AL ZOUR
Bids	2005	2006	2007	2008
Electricity ($ cents per kWh)	1.42	1.5	2.7	2.76
Water ($ cents per cubic meter)	57.07	82	102	110.35

Source: MEED, MEES

capital costs decreased significantly. In the US, CCGTs' capital costs declined from $ 600 per kW in 1990 to $ 400 in 1996. A similar decline has been noticed for coal-fired plants. Technological developments have certainly enabled price declines, as did pressure on equipment suppliers to cut costs. The escalation in costs after 2000 clouds the assessment of the impact of the introduction of competition in MENA countries. Indeed, in line with global trends, capital cost increases translated into higher electricity prices in projects' bids. The following table summarizes a few illustrative examples (Table 4.1):

This increase has been more marked in some countries than others. For example, between 2004 and 2008, Bahrain witnessed an increase of 75% in power tariffs.

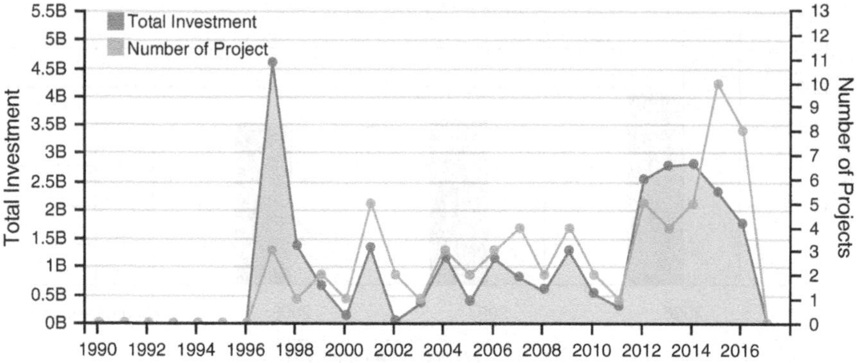

Fig. 4.6 Private investments in the electricity sector in MENA ($ US million). Source: World Bank's PPI (Private Participation in Infrastructure) Projects Database

4.2 Subcontracting Without Delegating: What Will Make a Reform Encourage or Hamper Needed Investments?

The reforms in the region were partly spurred by the increasing investment needs in the utilities sector. The following figure shows the amount of private investments in the electricity sector in the region between 1990 and 2016. The investments include management and lease contracts, concessions, divestitures and Greenfield projects. During 1996–2007, MENA received a total of $ 13 billion of private investments in its electricity sector, while $ 315 billion have been spent globally. MENA and Sub-Sahara Africa ($ 6 billion) were the two regions with the smallest private investment shares.

The World Bank (Private Participation in Infrastructure) Projects Database referenced in Fig. 4.6 lacks projects in large countries such as Saudi Arabia. Based on a non-exhaustive compilation of projects similar to the one presented in Appendix F a partial amount of investments injected in electric generation projects in MENA is estimated. This showed that investments have increased from at least $ 0.5 billion per year in the late 1990s to more than $ 4.5 billion in the late 2000s. This increase is driven by two factors:

(a) The demand growth detailed in Chap. 2 called for additional investments, mainly in Greenfield generation.
(b) The projects' costs escalation noticed in the whole energy sector is relevant in the electricity sector as well (escalation of EPC, labour, steel and other raw material costs in recent years) in addition to the tightness of the energy services sector.

The relatively high levels of investment between 2012 and 2015 were in reality concentrated in a single country. Morocco, with a series of coal, wind, and more notably solar (CSP) investments dominated electricity investments headlines during that period.

Electric reforms facilitated the introduction of private capital, but they were not sufficient in some cases as a few countries had to revert to EPC contracts to make sure that plants will be built and that projects will proceed. This contributes to an element of "moral hazard" taking root in the authorities energy policies—that is, tendering authorities could always decide to revert back to EPC contracts, which they can more easily control, if not enough interest in IPPs, instead of trying to improve the IPP terms. As a matter of fact, the MENA region has been in competition with other regions for investments from competent private companies. During the 1990–2007 timeframe, nearly $ 120 billion of private capital were invested in Latin America and the Caribbean, in comparison to the $ 13 billion invested in MENA.

Clearly, the lessons learnt from the region and elsewhere teach us that successful IPPs were built on an integrated set of criteria:

- Credibility: once the contractual structure established, counterparts—the electricity purchaser for instance—should be credit-worthy entities with world class financial/technical advisors and risks properly aligned. The sites and permitting process should be clearly defined. Timing must be respected as much as possible, and when that is not the case, it would be explained for predictable or external reasons.
- Clarity of processes: clear bidding processes, in order to attract the right type of players resulting from a sufficiently large cadre of companies/consortia in competition. Experiences in the region showed that, at times, local authorities had to "accept" the only bidder that they had even if he does not fully fulfil the conditions, or to re-tender the project. This could result in additional cost and delays for the evaluation process.[12] In general, it is even preferable to call in for independent review to define the terms of the evaluation as well as the project structure. The efficiency incentives, when they exist, should be clearly stated and reviewed; Egypt had to review its "internal power pool" as it was too opaque to enable generators to compete while Abu Dhabi was clear since the inception on the RPI-X performance criteria. In the same vein, government intervention on PPAs terms should be limited to driving performance incentives: end-of-year adjustment to bring RoE of all operating companies to uniformity (such as in Egypt) is a disincentive for performance. Transparency is also useful as IPPs can serve as performance benchmarks when they are under similar/comparable terms and conditions as government-owned utility assets.
- The fuel offer is an important element of IPPs' attractiveness, and relates to the issues of pricing, availability and policy orientations (governments' fuel mix policy choices). Most resource holder countries in the region, with largely

[12]For example, in 2005, the Saudi Water and Electricity Company (WEC) received only one bid for Shuaiba (900 MW/176 MMgd cogeneration plant) from a local/Malaysian consortium: Acwapower Projects, Tenaga Nasional Berhad and Malakoff Berhad. Some prequalifiers were concerned about investing in an oil-fired (and not gas) IWPP located on the west coast of Saudi Arabia; the others were not able to bid after WEC declined to extend the deadline.

Table 4.2 Saudi IPP Power Projects For Implementation: 2010–2017

Projects	Capacity (MW)	Investment (SR Mn)	IRR (%)
Muzahimiyah P/P	1725	2800	14.91
Rabigh-II P/P	2400	5000	15.79
Qurayyah-II P/P	3600	7200	16.47

available fuel, followed the "$ 1 gas rule"—gas, middle distillates or crude being provided at subsidized prices. However there has been increasing pressures to introduce new gas pricing mechanisms (at least cost plus, or market price). At the same time, fuel cost advantage is not always sufficient to attract enough of the right type of players, as the Iranian experience shows. Conversely, a country like Morocco with poor resource endowment but with poorly-subsidized tariffs and market-based reforms was able to attract IPPs. In general, if future fuel availability is a concern, operators will usually ask for a flexible mechanism or even a renegotiation clause: when the fuel supply agreement is challenged—security, volumes and/or pricing terms-, PPAs might need to be renegotiated.

- As far as the specific players are concerned, it should not come as a surprise, however, that despite the relative attractiveness of the region, there is a declining number of companies actively pursuing investments. As it is the case in most parts of the world, severe financial constraints have forced many western companies to retreat from foreign markets in order to refocus on core businesses at home, clean up their balance sheets and re-establish the trust of financial markets. However, it is interesting to point out that to date the companies with power projects under construction or in operation in MENA countries have not yet moved to sell-down their stakes in these investments. M&A deals in the electricity business also reduced the number of international players.

In general, rates of returns are not the only criteria that are relevant to a project's attractiveness: while Abu Dhabi offers 13% rate of return, Saudi projects were not exceeding 8–9%. Yet, the Saudi experience shows that companies could consider additional (qualitative) criteria—expectation to access upstream acreage—despite the lower rate of return. Subsequent Saudi IPPs IRRs are higher (Table 4.2).

- The PPA terms' risk profile reflected international project finance standards, and as such was reasonable from a risk point of view, particularly as PPAs were usually backed with sovereign guarantees from financially strong states. The foreign exchange risk on capital recovery portion may be an issue, but which can sometimes be mitigated by increasing the local component or using a mix of indices. A 15 year-PPA is on the shorter end of the range, 20 years is more typical, but it is extendable. The PPA length usually exceeds the payback period of the debt loans so that banks are not exposed to the offtake risk. International Arbitration in neutral country is preferred. Financial risks (foreign exchange, interest rate hedge, and price) should be hedged. Force Majeure [acts of man (strikes, terrorism), acts of nature (natural hazards, floods, earthquake), acts of government (embargo), impersonal acts (market freezes)] would require insurance and government support. Force Majeure can also result in renegotiations.[13] IPPs would take risk for insurable events. Captive IPPs for large industrial plants were at times easier to implement even in unfavourable environments for IPPs, as they bear a lower offtake risk. More conventional risk mitigation approaches include Take-or-pay/Take-and-pay clauses and marketing contracts.

There are other categories of risks that the various contracting parties have to take into account and the project viability would be tested against expert opinions and feasibility studies. Even with long-term supply contracts, with supply or pay penalties and price commitments, it is important to have explored alternative sources in order to mitigate fuel supply risk. Risks of projects cost escalation and delays would have to be tackled in the EPC contract (Competitive tender, fixed Price, date certain, delay penalties). The project's implementation should rely on a strong project management team and regular completion, performance and reliability tests.

Ideally, developers would take on risks upfront and governments would be more exposed over the long term and over uninsurable events. In the case of historical IPPs in MENA, governments assumed more risks than they needed to or bargained for.

Obviously, private investments are vulnerable to global developments like economic crises and liquidity shortages hampering availability of funds. Even so, the region was able to attract a portion of global private investments, although it could not attract as much as Latin America for instance. Moreover, several IPP/IWPP projects were switched back to EPC contracts because of insufficient bids or because of strained international finance. One can conclude that electric reforms contributed to attract some investments in the region but failed to cover all investment needs. For instance, the "fuel offer", embedded in long term "$ 1-gas" supply contracts, made the region attractive for power developers but it was not sufficient, and more importantly, it was not sustainable (see fuel price changes section in Chap. 3).

4.3 Summary

The first step in any reform evaluation is to analyse the effect on price for end-users. However, liberalization does not necessarily reduce electricity prices. In the case of directly subsidized systems, the only possible "price" indicator would be the wholesale cost. In countries which introduced competition in generation and implemented the single buyer model, wholesale tariffs or Bulk Supply Tariffs could be the best proxies to measure possible efficiency gains in the sector. The evaluation will also depend on the chosen method of regulation ("rate of return" or "RPI-X"). For example, Oman and Abu Dhabi apply the RPI-X method, while Jordan uses a Rate of Return method. The RPI-X was recognized as offering the best outcome

[13] See Sect. 3.2 of Chap. 3 for fuel price/volume renegotiations.

as it sets prices to be constantly reviewed downward and it encourages a 'good' quality of service to consumers.

The second step of this "common evaluation" consists in analysing whether reforms contributed to attracting private investments. The MENA region was able to attract a portion of global private investments, although not as much as other emerging regions (e.g. Latin America). Several IPP/IWPP projects were switched back to EPC contracts because of insufficient bids or because of strained international finance. Electric reforms contributed to attract some investments in the region but failed to cover all investment needs. The "fuel offer", embedded in long term "$ 1-gas" supply contracts, made the region relatively attractive for power developers in some cases, but it was not sufficient and not sustainable.

Chapter 5
Evaluation Step 2: How Optimal Is the Sequencing of the Reform?

Abstract MENA countries followed different reform paths in terms of the number of steps taken, the length of the process and the sequencing. A Herfindhal-Hirschman index (HHI) analysis shows that there are only two countries in the region where the dominant player is not the national utility or any other government-owned entity: Abu Dhabi and Morocco. In MENA, despite market concentration, wholesale prices are not volatile and kept at a relatively low level. However, increasing industry consolidation could translate into potentially higher wholesale tariffs, and higher subsidies. The dual market model of dual firms could be the response of developing countries to the "textbook model" as it was implemented in the UK or New ZealandSome countries are moving towards that pragmatic direction with the creation (re-creation) of national champions, staffed with highly educated technocrats, and partially publicly-listed. A strong regulation to prevent market abuse is key.

Keywords Market concentration · Wholesale prices · Dual firm · National champion

The World Bank (2002) defined the "optimal" temporal sequence of electric reform as: (a) launch a privatization/liberalization program, (b) enact an electricity law permitting unbundling and divestiture, (c) establish an independent regulatory authority, (d) approve a new power market structure, (e) unbundle the power utility, (f) privatize or close on a concession for some private distribution and (g) privatize some generation. Countries are not expected to follow all these steps, but most will take a combination of some of these steps and not necessarily in the same order. However, it is important to mention that IPPs' entry is generally perceived as a step "that might happen at any time in the sequence without being flawed if other steps had not been taken". It appears that this last point has been particularly followed by most countries. Table 5.1 shows the diversity of paths to reforming power sectors that have been taken by MENA countries. The paths are diverse in the number of steps taken, the length of the process and the sequencing.

L. Benali, *Electricity-sector Reforms in the MENA Region*, Perspectives on Development in the Middle East and North Africa (MENA) Region,
https://doi.org/10.1007/978-3-319-96268-9_5

Table 5.1 Reform steps and timings in MENA countries

	(a) Launch	(b) Law	(c) Regulator	(d) New market structure	(e) Unbundling	(f) Privatization–distribution	(g) Privatization–generation
Morocco	1994	1994	No	1997	historic	Partial	No
Algeria	1997*	2002	2005	1997	1997	No	No
Tunisia	1996	1996	No	1996	No	No	No
Libya	N.A.	N.A.	N.A.	N.A.	N.A.	N.A.	N.A.
Egypt	1984/1996	1996	1997**	1991	1983/1998/2001	Considered	Considered
Israel/Palestine	1996	1996	1996	2004	No	No	Considered
Jordan	1996	1996	2001***	1999	1999	2008	2007
Lebanon	2001	2002	No	N.A.	N.A.	N.A.	1994
Syria	N.A.	N.A.	N.A.	N.A.	N.A.	N.A.	N.A.
Saudi Arabia	2000	2001	2001	2000	Considered	No	No
Yemen	Long considered	2009	No	N.A.	N.A.	No	No
Oman	1994	2004	2004	1999	1999	Considered	Full
UAE (Abu Dhabi)	1997	1998	1998	1998	1998	Considered	1999
Qatar	2000	No	No	2000	No	No	Full but government retains shares
Bahrain	2002	No	No	2004/2007	No	No	2005
Kuwait	N.A.	N.A.	N.A.	N.A.	N.A.	N.A.	N.A.
Iran	2001	2002/2008****	No	No	No	Considered	Considered
Iraq	Considered	N.A.	N.A.	N.A.	N.A.	N.A.	N.A.

*Sonelgaz underwent a first restructuring in 1982 when it shed some ancillary construction and installation activities
**1st MD in 2001
***Commissioners appointed in 2005
****Iran's new Foreign Investment Promotion and Protection Act (FIPPA) was passed in 2002—with direct reference to BOTs for example- and its privatization law was ratified in July 2008 after several delays

This table generates some remarks:

(a) The time required to conduct a "complete" reform is very different among countries.
(b) Even if the spread of electric reforms in the region resemble a "domino effect", countries did not reform the same way.
(c) It is difficult to put an exact date on the launching of the reform. This time usually coincides with an official announcement or the drafting of a law.

Global experiences also reveal that the span of time required to carry out all these steps is extremely different from one country to the other. Indeed, World Bank (2002) cites a few enlightening examples: 1 or 2 years in Hungary (1994–1995) and Bolivia (1993–1995) to more than 9 years like in Brazil (1990–1999). Moreover, these same global experiences show that long reforms are not always the most successful (Brazil), although quick reforms bear the risk of being defective.

More particularly, presuming stability on various fronts over a long horizon (project's lifetime) is unrealistic. Bacon and Besant Jones (2000) argue that BOO/BOTs are suitable for large projects in stable financial and macro-economic environments, that they have many systemic prerequisites for success and that they should be avoided until power sector reforms are complete. Clearly, these conditions are difficult to achieve in developing countries, or in countries in transition. MENA countries are no exception to the rule. Their macro-economic situation is changing, progressive and sometimes widely fluctuating. Following scrupulously the World Bank's manuals, reform toolkits textbooks and ideal models might be challenging, and governments prefer to keep some flexibility in how they execute reforms when possible.

Most countries with liberalized markets have opened up their markets gradually. Those who introduced retail competition started with allowing the choice of supply for eligible consumers, large industrials, before they extended it to the residential sector. Competition delivered innovative tariffs arrangements, structures and demand management methods like smart metering. A fair assumption would be that this innovation would not have developed in the absence of competition (Littlechild, Joskow, Price).

By contrast, long reforms can be interpreted as the sign of rigid industrial structures and complex institutional set-ups (see section on attractiveness and feasibility of reforms). Another interpretation could be a high degree of market power. In those countries which have unbundled and privatized generation, it is possible to measure the degree of potential market power.[1] The Herfindhal-Hirschman Index

[1]Market power is possible when a single entity may gain too much control within the market and hinder the development of the marketplace. This may result in noncompetitive prices, inferior services, and slow development of new products and services. The concentration of market power, calculated as market share, is an indicator of the ability of a market participant to manipulate prices. Market power can be used to prevent new suppliers from entering the market or to eliminate existing competitors, not through the provision of superior products at lower costs, but through manipulation of the market. Firms with market power have the ability to extract excess profits from customers.

Table 5.2 Market concentration in MENA

	# of firms	Share of Largest firm (%)	HHI	Equivalent # of equal-sized firms
Morocco	3	43% (Jorf Lasfar)	0.3090	3.24
Algeria	3	75% (Sonelgaz)	0.9414	1.06
Tunisia	3	77% (STEG)	0.6326	1.58
Libya	Monopoly			
Egypt	4	87% (EEHC)	0.7628	1.31
Israel/Palestine		Quasi-monopoly		
Jordan	2	76% (CEPGO)	0.6226	1.61
Lebanon	4	90% (EDL)	0.8166	1.22
Syria	Monopoly			
Saudi Arabia	4	85% (SEC)	0.7324	1.37
Yemen		Quasi-monopoly		
Oman	5	EHC (56%)	0.3656	2.74
UAE (Abu Dhabi)	**5**	**23% (IPP-CMS)**	**0.1565**	**6.39**
Qatar	3	62% (QWEC)	0.4570	2.19
Bahrain	3	40% (Ministry)	0.3513	2.85
Kuwait	Monopoly			
Iran	Quasi-monopoly			
Iraq	Monopoly			

Source: local utilities and author's calculations

(HHI)[2] is the commonly used simple measure to assess this potential. HHI is the sum of the square of the market shares of the firms operating in the market. The index varies between the unity for a monopoly and zero for perfect competition—large number of equal-sized firms. Moreover, in deregulated markets, HHI can be used as an indicator of how much prices could be raised above the marginal costs of production.

Table 5.2 shows that there are only two countries in the region where the dominant player is not the national utility or any other government-owned entity: Abu Dhabi and Morocco. In Abu Dhabi, the diversity of players enables a fair

Some of the means that are employed include: (a) predatory pricing—lowering prices to eliminate competitors with the anticipation of subsequently rising prices; (b) cross subsidies—shifting costs to captive customers with relatively inelastic demand to lower prices to those customers with elastic demand; (c) tie-in arrangements—a sale in which the buyer, in order to get the item desired, must also purchase another item(s); (d) engaging in joint ventures that exclude competitors; and (e) controlling access to markets through delivery channels—limiting access between suppliers and customers in the transmission grid. The specific features of the electric sector create additional complications.

[2]The U.S. Department of Justice considers a market with a result of less than 1000 to be a competitive marketplace; a result of 1000–1800 to be a moderately concentrated marketplace; and a result of 1800 or greater to be a highly concentrated marketplace. As a general rule, mergers that increase the HHI by more than 100 points in concentrated markets raise antitrust concerns.

amount of competition, even if all have long-term PWPAs and even if TAQA has now a stake in all generation companies. In Morocco, Jorf Lasfar's market share used to exceed 60%.

Regulation by law appears to be the easiest mean to prevent the emergence of market power.[3] The Omani Sector Law for example provides for the implementation of market share, economic interest and regulated activity restrictions. The Regulatory Authority can relax these provisions if it considers it to be in the public interest. The Market Share thresholds (generation only) are not defined in the law, but the latter stipulates that each generation licence shall include a provision for the imposition of market share thresholds (25% of production capacity—although some cases might require to use the electricity generated). Barring exceptions (Transmission Company or regional companies for example), "no licensee shall be entitled to undertake more than one regulated activity or to hold any direct or indirect economic interest in any other License". These provisions are intended to hamper excessive concentration and possible market power abuse, and eventually to prevent vertical reintegration of the unbundled activities. As a reminder, unbundling has been mainly pursued in order to make industry costs transparent, and to gradually decrease long-term costs and subsidies, by encouraging competition.

The HHI analysis has its limitations since, in several cases, IPPs are joint-ventures where the "national champion" keeps an important stake (possibly more than 50%). Moreover, the long list of IPP and IWPP projects scheduled in the region would tend to indicate that market concentration will diminish over time.

In this regard, Jaime, Eduardo, Alejandro (2001) consider that in markets similar to the British type de-integrated model, market power by a small number of producers is not necessarily negative during period of capacity tensions, since this should create higher prices, which in turn stimulates the interest from new entrants and the development of new capacity.[4] Obviously, this will depend on the social acceptability of price peaks and market volatility as investment incentives.

In MENA, despite market concentration, the fact that the government keeps an indirect control over electricity generation, particularly through shares in the companies and long term PPAs, means that wholesale prices are not volatile and kept at a relatively low level. However, increasing industry consolidation and coalition could translate into higher prices in projects' bids, potentially higher wholesale tariffs, and higher subsidies should end-user prices left unchanged.

In MENA, countries reformed their electricity sectors very differently. Even if most chose the Single Buyer Model, they all chose different paths to reach it. The paths are diverse in the number of steps taken, the length of the process and the sequencing of the steps. Long reforms are not always the most successful but quick

---

[3]Ibrahim calls on experiences in Central and South America to argue that a regulation by law may be a temporary replacement to an independent regulator. He does not see it as sustainable solution given the dynamic nature of electricity markets. This temporary solution can however be leveraged to kick-start the reform process as a stepping stone to creating the independent regulatory environment needed within a reasonable period of time.

[4]Inter-American Bank specialists.

reforms bear the risk of being defective. Long reforms can be interpreted as the sign of rigid industrial structures and complex institutional set-ups, and might not solve the issue of market power.

The region displays an important concentration of players in the generation of electricity. That is probably at the expense of wholesale tariffs. Victor and Heller (2007) provide a theoretical framework to address this issue of concentration of players and market power in developing countries. The dual market model of dual firms is the response of developing countries to the "textbook model" as it was implemented in the UK or New Zealand. Dual firms can perform two distinct tasks simultaneously: they have the political connections to mobilize resources that are essential to their commercial viability (regular payment of subsidies, fuel supplies) and they are sufficiently well managed and have the necessary power to face attempts at being involved in uneconomic tasks (government social projects, excessive wage demands by the unions). Victor and Heller suggest that this outcome is likely and workably efficient in environments where market institutions are weak.

This framework could look complacent as it could be interpreted another excuse to avoid unbundling and competition. Dual market seeks to combine features of the "text book" reform with residues of State monopolies. However, when applied to MENA countries, it provides an attractive and feasible solution to the immediate concerns of the electric sector (performance, delivery and financial viability). Some countries are moving towards that pragmatic direction with the creation (re-creation) of national champions, staffed with highly educated technocrats, and partially publicly-listed. A strong regulation which would prevent market abuse (high wholesale tariffs, limited access to network) is key.

Today, economics cannot be narrowly defined as a social science concerned with the production and allocation of material goods anymore. Rather, it relates to incentives for several forms of social institutions. The evaluation grid developed in the next chapters could be applied to other types of reforms: social, macroeconomic, political, etc. It could particularly be used when benefits and losses are difficult to quantify. Evaluating electricity reforms in developing countries suffer from lack of transparent data. The absence of retail price signals because of subsidies can cloud the picture if wholesale costs are not accessible, and even there demand response is often overlooked and marginalized in analysis. In the absence of a direct link between generation costs and retail prices, the quantitative analysis attempted to analyse the impact of electric reforms in the MENA region on costs and investments. The impact on costs is unclear, but at least, one could conclude that electric reform did not lead to a substantial decrease in costs (comparable in the case of Abu Dhabi for instance to those of 1999) even if expectations of economic performance played a critical role in launching these reforms in the first place. The RPI-X regulation is acting as an efficiency incentive for most generators and government-owned assets will be gradually in competition with the others. The impact on investments is more clear-cut: electric reforms contributed to attract some investments in the region but failed to cover all investment needs. For instance, the "fuel offer", embedded in long term "$1-gas" supply contracts, made the region attractive for power developers but it was not sufficient, and more importantly, it was not sustainable. Future quantitative evaluations will be even more difficult as fuel prices might be reviewed and recent inflation of capital costs will come through the system. The role of institutions which can perform this evaluation exercise, ideally independent regulators but could be other entities becomes even more critical. That would assume however access to relevant data and information and availability of resources to conduct the evaluation.

5.1 Summary

MENA countries followed different reform paths in terms of the number of steps taken, the length of the process and the sequencing. Even if the spread of electric reforms in the region resemble a “domino effect”, countries did not reform the same way. A Herfindhal-Hirschman index (HHI) analysis shows that there are only two countries in the region where the dominant player is not the national utility or any other government-owned entity: Abu Dhabi and Morocco. In Abu Dhabi, the diversity of players could enable some competition, but all generation companies have long-term PPAs. In Morocco, Jorf Lasfar’s market share used to exceed 60%. Regulation by law appears to be the easiest mean to prevent the emergence of market power. In MENA, despite market concentration, the fact that the government keeps an indirect control over electricity generation, particularly through shares in the companies and long term PPAs, means that wholesale prices are not volatile and kept at a relatively low level. However, increasing industry consolidation and coalition could translate into higher prices in projects’ bids, potentially higher wholesale tariffs, and higher subsidies should end-user prices be left unchanged. The region displays an important concentration of players in the generation of electricity, probably at the expense of wholesale tariffs. The dual market model of dual firms could be the response of developing countries to the “textbook model” as it was implemented in the UK or New Zealand. It seeks to combine features of the “text book” reform with residues of State monopolies, and provides an attractive and feasible solution to the immediate concerns of the electric sector (performance, delivery and financial viability) in MENA. Some countries are moving towards that pragmatic direction with the creation (re-creation) of national champions, staffed with highly educated technocrats, and partially publicly-listed. A strong regulation to prevent market abuse is key.

quantitative valuations will be even more difficult as real prices might be depressed and recent inflation of capital costs will come through the system. The role of institutions which can perform this evaluation exercise, ideally independent regulators but could be other entities, becomes even more critical. That would assume [illegible]

5.4 Summary

[illegible]

Chapter 6
Evaluation Step 3: Attractiveness and Feasibility

Abstract The concept of attractiveness (economic), with the desirability (political), is central to the analysis of reforms in the neo-institutional economy. The impact on macro-economic policies (industrialization, economic diversification) is a key component of the attractiveness of MENA reforms. The more decentralized the institutional environment is, the less feasible the reform will be without negotiation and spending of political capital. The more rigid the industrial structure is, the longer the reform will be. Time is a critical factor as it increases the probability for the reform to be distorted, reversed or frozen. Transaction costs would be higher in countries with a rigid industrial structure, and it would be more difficult to reverse or to freeze a reform in an advanced stage. In a region where world-class oil and gas reserves are concentrated, the resource endowment of a country impacts the feasibility of a reform, and its process.

Keywords Attractiveness · Desirability · Industrial structure · Transaction costs · Resource endowment

As mentioned in the introduction, Finon (1997) and Hunt (2002) described in their research the two main factors justifying the existence in the past of industrial organizations in the electricity sector based on vertical and horizontal integration. The strong need for technical coordination between the different parts of the chain,[1] in addition to the "natural monopoly" nature of transmission, explained the requirement to bundle the generation and the transmission within vertical monopolies,

[1]Needs of coordination concern the long term (generation, transmission and distribution planning, demand forecast), the mid term (maintenance coordination, demand forecast, power reserve announcement), the short term (production offers, determination of energy quantities demanded and offered, evaluation of ancillary services availability, production plan setting without discrimination), even occur in real time (measuring and pricing of difference between previsions and realized transits, frequency and voltage regulation, emergency situations), or even ex post (balance of payments and clearance of dues). The time scale can range for tens of years in the long term down to the micro-second in real time.

L. Benali, *Electricity-sector Reforms in the MENA Region*, Perspectives on Development in the Middle East and North Africa (MENA) Region,
https://doi.org/10.1007/978-3-319-96268-9_6

sometimes with the distribution part as well. Transactions and cost allocations between the different parts were also difficult to measure, reinforcing the need for a continued internalized process and centralized institution.

The somewhat loose, but nonetheless common, concept of national or general interest created the need to have these centralized institutions not only under the tutorship of the State, but also commanded and regulated by the State. The "natural monopoly" nature of the transmission has long been debated. Joskow and Tirole (2003) conclude that a market-driven investment framework is likely to lead to inefficient investment decisions in electricity transmission. The design of regulatory mechanisms that would stimulate efficient investments by regulated incumbent transmission owners and by merchant owners is very challenging.

More generally, when the State is a strong and institutionalized entity, and when funding is not an issue, the objective of building capital-intensive assets in a timely, sometimes accelerated, manner is usually fulfilled. However, Joskow (1989, 1991) demonstrates that the integrated monopoly:

- Lacks mechanisms to incentivize efficiency
- Can not easily adapt to technological progress and innovation
- Does not guarantee optimized production costs

Chapter 2 clarified the main reasons why MENA countries moved towards reforming their electricity sectors. In this section, these reforms will be tested against a constructed criteria grid, built on the concepts of Attractiveness, Feasibility, Credibility and Remediability.

6.1 Attractiveness: What the State Needs, the Single Buyer Model Provides

Developed in World Bank (1995) and Menard and Shirley (2000), the concept of attractiveness (economic), with the desirability (political), is central to the analysis of reforms in the neo-institutional economy. According to this framework, the attractiveness and desirability of a reform are usually driven by three main factors: an economic crisis, a sectorial crisis or a major political change. However, this theory, relevant in the case of reform leaders like the UK, disregards the impact of "exogenous factors". It has already shown its limits in the case of most European countries which simply had to follow the European Directive 96/92 CE. It appears even less relevant in the case of most countries in the Middle East and North Africa. As explained in Chap. 2, all the countries in the region which introduced electric reforms were not driven neither by a sectorial crisis nor by a major political change. In some countries, economic crisis might have probably been the most relevant factors in the pre-2000 period, but surely not afterwards. The history of electricity reforms in the region shows that the attractiveness and desirability was mainly driven by other factors which do not fall necessarily in these three categories. Depending on the country, most major electricity reforms in the region were mainly driven by the international "vogue", and by the need to engage in a different dynamic. Expectations of economic performance, while benefiting from some of the lessons learnt from other international experiences, played a critical role (Chap. 2).

In some very specific cases, reform was driven by the prerogatives of international lending institutions. This does not mean that privatization revenue and proceeds were not an important motivation, particularly in some very specific cases of hydrocarbon-importing countries or energy exporters during periods of budget constraints. However, the history of reforms in the region shows that in all cases, the focus has tended to be on the introduction of competition and that exogenous factors, such as international trends, were important in that process.

Another aspect of reform attractiveness, often disregarded, is the impact on macro-economic policies. As explained in Chap. 2, the desired industrialization contributed heavily to the growth in electricity demand in the region, at least on a baseload basis. It also contributed, in countries with monopolies structures, to launch the debate on the need to attract foreign investors in electricity generation for captive power plants. In these cases, the need to reform the electricity sector stemmed from the "opportunistic" need to prepare the regulatory framework in order to allow private generators to operate in the country.

Attractiveness of reforms relates to the anticipation of gains or losses that different parties (shareholders, workers, new entrants, consumers and governments) with negotiating power would experience while political desirability relates to the political advantages and constraints faced by governments. It is important to differentiate the ex-ante anticipation of gains/losses from the ex-post evaluation of these gains/losses. Attractiveness is conditioned by the former.

One would assume that the authorities who initiate reform programs would always emerge as the winners, but World Bank (1995) reminds us that "*the reform can cost a government a support base, because reforms almost invariably involve eliminating jobs and cutting long established subsidies*". A reform remains attractive for the initiators (in most cases the government) so long as it does not cost them a support base. This is another reason why reforms in the region were generally light-handed, preserving jobs and rarely resulting in a change in tariffs (and even when tariffs were changed, other reasons were used—international energy prices, DSM programs). In most cases, the government could not be considered as having emerged as a winner since the political responsibility of managing the electricity sector was not entirely transferred to independent entities. Even in areas where private companies could be blamed for failures or blackouts, the government is the one held accountable for having allowed the entrance of these private companies. Moreover, the Abu Dhabi and Oman experiences show that reform did not automatically result in a sizeable decrease in production cost, i.e. no sizeable decrease in the volume of subsidies which would offset the cost of the reform was evident.

Usually, various analyses of reforms' attractiveness depict the perception of gains and losses of the different parties (shareholders, workers, new entrants, consumers and governments) before the reform. In the case of MENA, the shareholder was usually the State itself, the workers did not have a say (except in rare instances where unions are relatively powerful like in Algeria), the new entrants were usually foreign companies willing to engage in competitive bids (no local private sector to invest in electricity) and the consumers were insulated by highly subsidized tariffs that they were promised to keep.

Would the government *necessarily* emerge as a winner from the process of electric reform in MENA? In order to answer this question, looking at the various identities of the State and the change of these identities before and at the launch of the reform process will be helpful. A theoretical construction frequently used by experts of Asian economies and some other developing economies will be used. This model is based on the opposition between two, interdependent but sometimes mixed, roles played by the State: the *"Clientelist State"* and the *"Developer state"*.[2] In a nutshell, the Developer State, economically-driven, provides the necessary resources to the Clientelist State. The latter, politically-driven, provides the social and political stability which is necessary for the well functioning of the former. C. Sautter,[3] while describing Japan, refers to a "pro" state (producer, protector, prospective and programmer—simultaneously or successively). In most countries in the region which launched an electric reform process, this occurred when the two (or several) identities coincided. The Clientelist State, during the days of government-owned monopolies, had to provide jobs for the local population and subsidized electricity and water for consuming citizens. The Developer State is in charge of producing electricity and water, but also of investing in infrastructure, education, health and ensuring that natural resources are properly dealt with. Barring exceptions, the reform occurred when the State (Developer and Clientelist) realized that it could continue to perform its dual duties without penalizing any one of the two. In this regard, *the State would need a sector structure that would enable to discharge it from the role of primary investor and electricity producer while maintaining the option to intervene to induce investments.* IPPs introduction and the Single Buyer Model allow that.

Drawing from the regional experiences, it appears that the government could be considered as a winner with the conjunction of these three aspects:

- Even without a change in tariffs, the introduction of competition is supposed to decrease costs and improve performance and therefore to decrease the amount of general subsidies.
- The "political" responsibility for the sector is transferred to other stakeholders. For example, blackouts, technical problems, even tariff increases (distribution) could be "blamed" on the new entities.
- The attractiveness of a gradual increase of fuel price to electricity generation, particularly in case of future higher marginal fuel supplies (as it would usually be the case), would gain in importance. This increase would ensure that natural

[2]BOUISSOU, Jean-Marie, L'économie politique de dix ans de crise, *Economie Internationale*, n°84, 4ème trimestre 2000.

[3]C. SAUTTER, Japon, *le prix de la puissance*, Paris, Seuil, 1973.

resources are adequately valued and generators would be encouraged to seek higher levels of fuel efficiency.

Earlier analysis of the BST indicates that the first condition is not fulfilled. On the contrary, the BST increased in some instances as inflation of EPC costs appeared in generation costs over the next few years. In the longer term, as concessions' expiry approaches and gas turbines (shortest lifetime among plants' equipment) need to be changed, investments and expenses will pass-through to the terms of the new purchasing agreements. However, this increase is due to exogenous factors, which would have occurred regardless of the reform programs. The government would assume that the introduction of competition and competitive bids would guarantee lower costs than if it had to go it alone. Therefore, deregulation would not necessarily decrease subsidies (if tariffs are unchanged), but it might guarantee levels of cost which are close to the costs achieved in perfect competition. The economic benefits would be more visible in the mid-long term.

The second point, relating to the political responsibility, occurred at some occasions, but the outcome for the government is mixed. Examples include when captive power plants or major IPPs were blamed for shortfalls, brownouts or blackouts. But ultimately, since in most cases the government did not disengage fully from the sector (or at least not yet as in Oman), it could be viewed as accountable for potential operational problems. This will depend on the public's perception of this accountability. The potential for tariff increases is interesting and its dynamic is complex. The government can not easily use the reform as a justification for retail tariff increase if any, as the government was the initiator and is assumed to keep a direct stake throughout the process. However, as explained by Abdullah Saif al Nuaimi, the reform "would help change mentality".[4] Here, in countries where a social pact between the ruler and the citizens is built on the free or heavily subsidized provision of utilities and energy, the attractiveness of the reform would relate to its long-term potential to alter that social pact.

Finally, the third aspect is the most pressing, given the increase in fuel supply costs or the starting of energy imports in some producing countries. The "fuel offer", embedded in long term "$1-gas" supply contracts, made the region attractive for power developers but it was not sufficient, and more importantly, it was not sustainable. The main outcome of the reform would be that the government would have built pure commercial relationships with the generating companies. Therefore, offering higher fuel tariffs or renegotiating fuel supply agreements would be probably more feasible, albeit not necessarily more acceptable, than if the generating companies were government-owned.

It is generally after the reform (particularly long ones) that the analysis of attractiveness (of further reforms) becomes academically interesting. Indeed, new players emerge and the government is neither the sole prime-mover nor the sole stakeholder. The relatively long and "one step forward two steps back" process in

---

[4] See Middle East Economic Digest (MEED) January 2008.

Saudi Arabia is particularly interesting in this regard. Saudi Arabia set up in 2003 the Water and Electricity Company (WEC), by royal decree, to act as the single buyer from the country's IWPPs.[5] A more detailed description of the reform of the Saudi electric sector is provided below in the feasibility sub-section. Saudi Arabia launched its electric reform relatively late given its weight in the GCC. Today's set-up, with different responsibilities between the various players (SEC, WEC and SWCC) and the regulator ECRA clearly impacts the attractiveness of the remaining steps in the process of reform, particularly during a period of tight finance and tough international conditions for utilities.

6.2 Feasibility: The Critical Role of the Institutional Set-Up, the Industrial Design and the Resource Endowment

The **feasibility** relates partly to the institutional endowment, in other words to the formal and informal institutions (as defined by North). This research deals with the most important formal institutions to economic analysis. In the same vein, the neo-institutional economy deems it important to look at (if and whenever applicable):

- The relations between executive and legislative powers in scope and intensity
- The ex-post constitutional control of laws by the court
- The organization of political parties and the electoral rules
- The role of the judicial power in economic regulation

In order to capture all these elements, two levels of feasibility are distinguished. First, the institutional feasibility depends on the degree of centralization of powers and decision-making. And second, the industrial feasibility "measures" how easily transformable the industrial structure could be (Table 6.1).

In the region, Kuwait would be the extreme example where the feasibility of electric reforms would be dependant on a strong and tough negotiation between the different players. In Kuwait, the role of the Parliament is critical and reform plans usually get caught in the middle in the strained relationship between the Parliament on one side, and the Government, the Oil/Energy ministry (and the Kuwait Petroleum Company) on the other. Following the 2003 parliamentary elections, the Emir of Kuwait, Sheikh Jaber Ahmad al-Sabah, appointed on 14 July Kuwait's 21st cabinet since independence in 1961. The most significant move has been the separation of the post of Crown Prince from that of Prime Minister.[6] It is the first time that the two positions were split. Sheikh Sabah has been widely recognised as the *de facto* Prime Minister of Kuwait and has been effectively running the

[5]WEC buys electricity and water from IWPPs, but sells the water to the Saline Water Conversion Company (SWCC).

[6]Crown prince Sheikh Saad al-Abdullah al-Salem al-Sabah gave up his inherited role of Prime Minister, and Sheikh Sabah al-Ahmad al-Jaber al-Sabah, who had been hitherto First Deputy Prime Minister and Minister of Foreign Affairs, has been appointed Prime Minister.

Table 6.1 Different forms of feasibility mix (institutional and industrial)—examples

		Institutional feasibility		
		Centralized institutional environment	Mix/hybrid environment	Decentralised institutional environment
Industrial feasibility	Flexible industrial structure (easily transformable)	Oman	Morocco	Kuwait
	Rigid industrial structure (non-easily transformable)	Saudi Arabia	UAE (federal)	Iran
		Vertically imposed (absence/weakness of parliament/legislative powers)	**Negotiated** reforms and **partially imposed**	**Negotiated with high constraints of participation**

government because of Crown Prince Sheikh Saad's illness. *The splitting of the two positions means that the Prime Minister is henceforth questionable by Parliament.* In a less publicized move, the Ministry of Oil and the Ministry of Electricity and Water were merged into a single Ministry of Energy.

As a result of the formation of a new cabinet, the Government issued a decree to reconstitute the Supreme Petroleum Council (SPC), the body which has the last say on strategic energy decisions. It confirmed then the new Prime Minister in its former position of SPC chairman, by stating that the *ex officio* SPC Chairman would be the Prime, and noted the merger of the Oil and Electricity/Water ministries. Consequently, apart from the Chairman, the new SPC consists of six members (the minister of communication/planning, energy, finance, trade/industry, the foreign minister and the Governor of the Central Bank of Kuwait). But these changes did not break the political stalemate over major energy reforms (mainly in the oil and electricity sectors). In March 2008, when the Emir of Kuwait, Sheikh Sabah al-Ahmad al-Jaber al-Sabah, dissolved parliament and called for the second general elections in less than 2 years, he hoped for confirmation of the government's actions. Despite the energy of KPC's new CEO, Saad al Shuwaib, and the conservative credentials of the minister of Energy, Mohammad al Ulaim, it appears that the success of various energy reforms depends mainly on the Prime Minister's ability (and will) to use and spend political capital. The Members of Parliament have been continuing since then to question the cabinet and to scrutinize energy deals. As a result of the 2008 elections, the parliament veered a little more towards Islamists, both Sunnis and Shi'a, but it did not mean they are aligned, and there was no clear fault line within parliament. Actually, Shi'a MPs side often with liberals on economic issues. However, tribal MPs were considered as the natural support of the cabinet.

The strained relations between the Parliament, the cabinet and KPC have several implications for any electricity reform. Contentious energy files, particularly those linked to the introduction of foreign companies in the energy sector (like Project

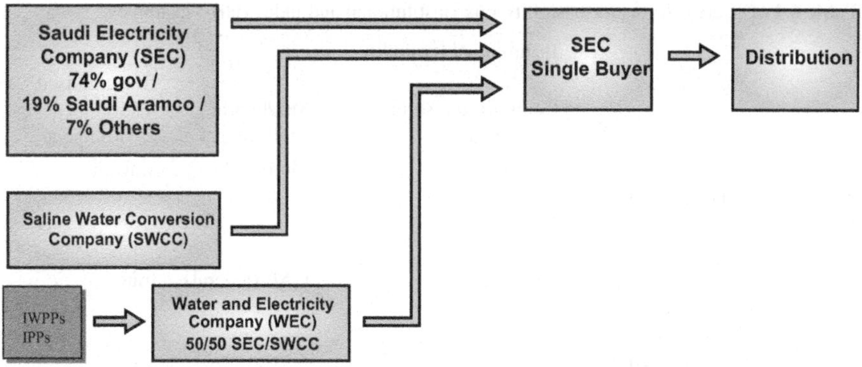

Fig. 6.1 The structure of the electricity sector in Saudi Arabia

Kuwait, the Al Zour refinery or the petrochemical Joint Venture with Dow), have been constantly distracting the Parliament from electricity reforms.[7] Moreover, delays on critical fuel supply projects (refining, upstream oil, development of non-associated gas reserves or LNG imports) jeopardize further the possibility to cover the power sector needs.

The first conclusion from this feasibility evaluation is that the more decentralized the institutional environment is, the less feasible the reform will be without negotiation and spending of political capital. The industrial structure is also an important aspect of the feasibility. It explains for example the relatively slow pace of market restructuring in Saudi Arabia (Fig. 6.1).

In Saudi Arabia, 85% of the Generation in addition to the Transmission and the Distribution is the responsibility of the Saudi Electricity Company (SEC). SEC resulted from the fusion in 2000 of ten regional entities. The transmission part of

[7]Since 1998, KPC has been arguing that it needs foreign expertise, particularly to solve reservoir management issues and corrosion problems. However, it had to face strong opposition from Parliament, on the ground of surrendering national sovereignty on hydrocarbons' reserves to foreign firms. Project Kuwait was the first of a series of projects designed to involve foreign oil companies. It aimed at increasing the northern fields' capacity from 600,000 b/d to 900,000 b/d within a time span of 5 years, as part of a strategy to double the overall capacity to 5 million b/d by 2020. The project was originally expected to cost some $7 billion, but KPC, while awaiting the IOC's proposals, envisions the total value to fall within a $2–$7 billion band. Given the legislators' reluctance over a first draft law that would enable the project to start, the Kuwaiti government has submitted in January 2003 a modified version to the National Assembly. The law answered some of the Parliament's concerns, as it specifically rules out the possibility of production sharing contracts, restricts foreign oil companies' involvement to technical support and operating agreements, and bans them from using local sponsors. The main stumbling block remained on the legal side: KPC along with the oil ministry wanted the Parliament to pass an enabling law and to approve a general draft contract's term sheets, thus preserving KPC's freedom to conclude directly with IOCs any specific contract. However, the National Assembly has shown strong reluctance to give up on its previous requirement to be involved in the decision-making process by considering, and approving, the terms on a contract-by-contract basis.

SEC was unbundled and transformed into a separate and autonomous entity in 2005. The other major entities active in the electricity sector are:

- **SWCC**: created in 1974, state-owned entity (affiliated to the Ministry of Electricity and Water) in charge of national production of desalinated water, produces electricity in its cogeneration plants for its own needs, and supplies SEC with the surplus.
- **Marafiq**: State-owned entity responsible for supplying power/water to companies in Jubail and Yanbu industrial zones.
- **WEC**: 50/50 owned by SEC and SWCC created in 2001 to support IWPP projects
- **ECRA**: Electricity & Cogeneration Regulatory Authority, created in 2001[8]

The State participation in IWPP projects is fixed at 40% (32% from Public Investment Fund and 8% from SEC). WEC is the single buyer of output of these IWPPs and it sells water to SWCC and power to SEC.[9]

Even if effective reform started in 2001 with the creation of WEC and ECRA, the Saudi Electricity Law was only published in 2005 and its implementing regulations in 2007. Arthur D. Little prepared the Masterplan for the restructuring of the sector and the landmark agreement to establish the country's first IPP (at Jubail for Saudi Petrochemical Company) was signed in 2003. Important considerations were included in studies for the development of an Institutional Plan with an Organizational Structure and an Incentives Program for Private Sector Participation, or the costing of electricity generation, tariff setting and structure. These studies were feeding ECRA's recommendations on these issues, and led to a long-term plan for electricity and desalinated water.

ECRA and the Saudi authorities have been clear about the future steps needed in the electricity sector. These include (in order):

- establishing an appropriate legal and regulatory framework, and restructuring the industry,
- regulating natural monopolies (transmission and perhaps distribution networks),
- corporatizing and/or privatizing existing assets and promotion of new private sector investments,
- allowing for cost-reflective electricity tariffs of regulated activities and creates a wholesale market for competition in generation, with a principal buyer.

In 2012, SEC provided an update of its unbundling process, into SEC and six fully owned subsidiaries (four generation companies (GENCO's), one Transmission

[8]Licenses for existing electrical operations (SEC) have been issued by Regulator (ECRA) and for new operations to Jubail Energy Company (JEC), the captive IPP to SADAF, to the WEC (trading electricity and water) and to Tuhama Energy Generating Company (4 IPPs for supply to Saudi Aramco). Some large consumers are also licensed to undertake electricity/co-generation activities (Al Obican Industrial Investments, Yanbu Cement Co., Yamamah Cement Co., National Prawns Co. (AlRubian)).

[9]There are however two exceptions: IWPP in Jubail monitored by Marafiq and the captive IPP for SADAF—in Jubail as well—which is the first Saudi BOOT.

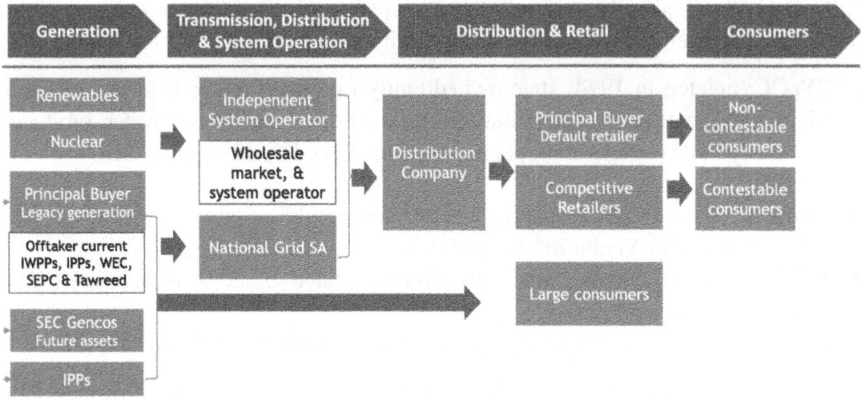

Fig. 6.2 Bis: end-state competitive electricity market (CEM)

Company (National Grid), and one Distribution Company (DISTCO). The preparation of SEC's unbundling program started in 2007, with the set up approved in 2010. The plan was for a restructured SEC with the National grid to go live in January 2012 and the four GENCO's with the DISTCO to be set up in January 2014. This was in line with the current transformation of the Saudi electricity sector from the single buyer model, to the "principle" buyer, and then to the open market (Fig. 6.2).

The competitive electricity market (CEM) roadmap was developed by the Electricity and Cogeneration Regulatory Authority (ECRA) to guide the process plans to move from SEC's unbundling to new fuel arrangements and the establishment of a wholesale energy market (with a full spot market for energy and ancillary services and bilateral contracts between generators and large customers). The view is to ultimately introduce a capacity market in the framework of a fully competitive market.

However, the final end-state is still subject to review and government decision. As of 2018, the transmission company and the Principal Buyer were created and operational but progress on the rest of the plan has stalled. It is interesting to note that the first step of the evolution proposed is similar to the attractive, feasible and credible model proposed at the end of this book.

The size of the Saudi market, the "hybrid" industrial structure of the electricity sector, the implication of several stakeholders including but not limited to Saudi Aramco (fuel provider, important power generator and an SEC's shareholder), made it difficult to reform the sector in a short period of time, compared to other MENA countries with smaller markets and simpler institutional set-ups. This could increase the probability of delaying the reform further.

However, in April 2016, the government released a blueprint of economic and social reforms called Vision 2030, driven by the Deputy Crown Prince Mohamed Bin Salman. The Vision intends to maximize the country's natural advantages in low-cost energy, substantial mining resources and availability of capital to accelerate the transformation and the diversification of the economy. The unfolding of Vision

2030 will have important ramifications for many sectors, including the electricity sector, and could provide the necessary momentum to push the sector's reform in the next few years.

The Saudi case illustrates how a long reform can continue to be alive and sustained for many years, for three main reasons:

- The reform remained high on the government's agenda, facilitated by a centralized institutional environment,
- The transaction costs incurred by the transformation of its industrial structure seem to outweigh the potential benefits of stopping the reform, and finally,
- The reform of the electricity sector coincided with a period of economic prosperity in Saudi Arabia (budget surpluses and a new dynamic of government's investments)

The second conclusion from this feasibility evaluation is that the more rigid the industrial structure is, the longer the reform will be. Time is a critical factor as it increases the probability for the reform to be distorted, reversed or frozen, sometimes because of external factors (e.g. global recession). However, this statement is mitigated by transaction costs which will be higher in countries with a rigid industrial structure. In other words, in these cases, it is more difficult to reverse or to freeze a reform when it has reached an advanced stage.

In a region where world-class oil and gas reserves are concentrated, a feasibility analysis would not have been complete if it did not look at how the resource endowment of a country affects the feasibility of a reform, and its process. Most theoretical and empirical studies on the reform of electricity sector and network industries seek to establish a relationship between political, ownership/institutional indicators (as explanatory variables) on the one hand and the reform process, the system's efficiency, the quality of supply, reliability, performance and investment incentives (as results)[10] on the other hand.

The rest of this section complement these analyses by investigating whether specific features of the country, like the resource endowment which influences the fuel mix, would have an impact on the reform process and results.

When the electricity fuel mix is dominated by a few power plants with high degree of interdependence, reform becomes more difficult. In general, the "indivisibility of capital-intensive assets" (nuclear reactors, hydropower dams) makes privatization and divestment difficult (Newbery 2002; Finon 2001), particularly if the size of the individual plants is large enough to dominate the market.[11] Actually, it is generally accepted that the "competitive paradigm" emerged because of the development of small-scale, highly-efficient power plants, and moeusually equipped with CCGTs.

[10]Wallsten (2001a) applies that relation in the case of the liberalisation of telecommunications in Latin America and Africa, Zhang and Kirkpatrick (2002)—for automous regulation-, Zelner and Henisz (2000)—for credible democratic institutions—and Steiner (2000)—for vertical unbundling—as ways to induce performance in terms of prices, investment and capacity utilization.

[11]As demonstrated by the success of Nordpool, hydropower is not necessarily an obstacle to market reform (as long as the size of the plants does not dominate the market).

Nuclear power in particular has a "special status" given operational, performance and waste management risks. For example, the French nuclear industry received preferential treatment and government aid to develop. And the preferential treatment could be extended even under competitive market conditions. A recent example occurred in 2003 in the UK where low wholesale prices obliged several thermal IPP companies to go bankrupt while British Energy received state aid and survived.

As the fuel mix appears to have a direct impact on the feasibility of a reform, the resource endowment of a country, which directly impacts its fuel mix, could influence the reform process and outcome. Once again, the case of Saudi Arabia is particularly interesting for three main reasons:

- Saudi Arabia has the world's largest reserves of oil (264 billion barrels, 25% of global proven reserves) and 7570 BCM reserves of gas.
- It is a major oil producer (ranks first or second globally) and exporter and a critical OPEC player.
- The size of the Saudi electric market is substantial in its own right. This last point is important as it appears that in small countries in terms of electric generation, dependant on a small number of power plants, electric reform might fail to introduce competition as it would create an oligopoly (Milan and Vives 2001; Ruffin 2002; Jamasb 2002). Woo's analysis (2002, 2003) of reforms in Israel and Hong Kong shows that "market reform is unlikely to succeed in a small market without interconnection" and "is almost certain to fail in a country with rapid growth (of electricity demand), limited generation sites, and no indigenous fuel".

Saudi Arabia, because of these three reasons, is also unique in the way its resource endowment have influenced the reform of its electricity sector during the first decade of this century.

Saudi Arabia made it clear that natural gas was not to be used for export, but for the domestic market, and particularly as fuel or feedstock for industrial development (e.g. petrochemicals). Gas-fired generation accounted for less than a third of power generation, the rest being oil-fired. Today, gas is considered as the fuel of choice for power generation but in the early 2000s, aluminium smelters, which would use gas for power generation, were encouraged to use crude.

At the same time, two-thirds of Saudi gas reserves are associated—with oil. Because of the long history of oil predominance in the energy industry, in the early 2010s, only 15% of the Kingdom has been reportedly "adequately explored for gas".

Launched in 1998 by then Crown Prince Abdullah, the Saudi Gas Initiative surprised the international community and led to hopes that oil would be included in the opening of the sector. The move was deemed also partly driven by political reasons.[12] The Gas Initiative was presented as a way to face the growing demand for basic needs, water and power, to develop a gas-fuelled industrial base, to create job opportunities for the large number of entrants to the market every year and to cut

[12]At a time when Saudi-US relations were strained, following the bombing of the Khobar towers military complex (June 1996, 19 USAF servicemen killed).

Saudi capital expenditure and to divert funds to more socially-related projects. More importantly, it was designed to integrate upstream gas exploration and development and downstream projects (new power stations, water desalination plants and petrochemicals).

In 2002, the Supreme Petroleum Council announced the selected IOCs to form consortiums. In its original framework, the $25-billion project was built around three Core Ventures (CV). **CV 1**[13] (South Ghawar) consisted in ethane and NGL recovery for the Hawiyah Gas processing plant, in the development of gas supplies in South Ghawar and exploration in the Rub' Al Khali area, and in the construction of two 2000 MW power plants/desalination units (Jubail and Rabigh), and a petrochemical facility (Jubail and possibly Yanbu). CV 2[14] (Red Sea) calls for the exploration of nine offshore blocks, the development of the Midyan (onshore) and Barqan (offshore) fields, the construction of a pipeline to Tabuk and Yanbu, a processing plant, a petrochemical facility at Yanbu/Rabigh and the development of power/desalination projects at Dhuba. CV 3[15] (Shaybah) was anchored on gas exploration in the Rub' Al Khali area, development the Kidan and Shaibah gas fields, construction of a pipeline from Kidan and Shaybah to the Hawiyah processing plant and the construction of a 1100 MW power plant, a petrochemical unit and a desalination plant (75 mgd) with the supply of dry gas to the Uqair power/desalination plant and the supply of ethane and NGLs to petrochemical facilities in Jubail and/or Yanbu.

The negotiations were long and protracted. One stumbling block related to electricity generation was the projects' Rates of Return. Even though increases were proposed, through adjustments to the fiscal regime, to a point about midway between the initial proposal and high single digits (typically 8–9% for Saudi IWPPs projects), this was still perceived low compared to the yields of around 13% yield in neighbouring countries' IPPs and the 15–20% typically requested by IOCs in upstream projects.

The different expectations on several aspects resulted in a gap of understanding and an ultimatum was sent in May 2003. In June of the same year, CV 1 was terminated. The Saudi Gas Initiative in its original form was shelved but it was redesigned and relaunched a few months later by disassociating the upstream and downstream components and tendering them separately. Subsequently, Saudi Aramco made several new discoveries, reflecting the Kingdom's focus on finding non-associated gas (Karan Khuff gas deposit in 2006, Jana-6 find in 2007 and three gas discoveries in 2008—Rabib-1, Arabiyah-1, Hasbah-16).[16]

Today, the Saudi fuel mix and the energy policy are quickly moving, as the country embarked on a drive to implement market-based fuel and energy prices. In the past, there were at times strained relations between Saudi Aramco and SEC because of gas supplies,

---

[13]$15 billion; led by ExxonMobil (35%), Shell (25%), BP (25%), ConocoPhillips (15%).

[14]$5 billion; led by ExxonMobil (60%), Oxy (20%), Marathon (20%).

[15]$5 billion; led by Shell (40%), Total (30%), ConocoPhillips (30%).

[16]Saudi Aramco's 2009–2013 plan shows exploration spending hitting $990 million per year, up from around $600 million between 2005 and 2007. And these figures did not include Aramco's 20–50% investment in the JVs exploring for Gas in the Empty Quarter.

but utilities and industries alike have been benefitting from below-market gas prices of $0.50 and then $0.75 per MMBTU. The country increased gas prices to $1.25 per MMbtu in December 2015; a move to market-based energy prices, including for gas, is part of the transformation plan of Vision 2030.

Weinmann and Bunn (2004) confirm that there is a direct link between the resource endowment of a country and the reform process it would undertake. For example, one of their main conclusions is that "a country will delay or not undertake a substantial reform in its electricity sector if the energy autarky if that country is high".[17] This conclusion is based on statistical analysis of Latin American markets energy resource endowment, exports and consumption and electricity prices.

6.3 Summary

The concept of attractiveness (economic), with the desirability (political), is central to the analysis of reforms in the neo-institutional economy. The attractiveness and desirability of a reform are usually driven by three main factors: an economic crisis, a sectorial crisis or a major political change. However, this theory does not appear relevant in the case of most MENA countries. The introduction of competition and exogenous factors (international trends...) were important in that process. The impact on macro-economic policies (sought industrialization, economic diversification) is a key component of the attractiveness of MENA reforms.

The institutional feasibility relates respectively to the degree of centralization of powers and decision-making. The industrial feasibility "measures" how easily transformable the industrial structure could be. The more decentralized the institutional environment is, the less feasible the reform will be without negotiation and spending of political capital. The more rigid the industrial structure is, the longer the reform will be. Time is a critical factor as it increases the probability for the reform to be distorted, reversed or frozen. However, transaction costs would be higher in countries with a rigid industrial structure, and it would be more difficult to reverse or to freeze a reform in an advanced stage. In a region where world-class oil and gas reserves are concentrated, a feasibility analysis would not have been complete if it did not look at how the resource endowment of a country affects the feasibility of a reform, and its process.

[17]The energy autarky is measured by the energy exports minus the energy imports, divided by the total energy consumption.

Chapter 7
Evaluation Step 4: Credibility

Abstract The credibility of the reform—or its capacity to provide stability of commitments, and the ability of its governance structure to solve regulatory problems—is critically supported by the design of "effective" regulatory institutions. In MENA, clarity of roles, participation of relevant stakeholders, and accountability are usually guaranteed. The autonomy from political intervention and the predictability of decisions are less certain. The development of robust regulatory frameworks and strong institutions was hampered by under-funding and a reluctance of authorities to transfer independent decision-making to regulatory authorities.

The growth in the volume of subsidies is usually not offset with the hypothetical decrease in production costs due to efficiency and competition. A subsidy applied through fuel pricing can potentially be counter-productive to stimulating efficient use of energy. The restriction of fuel supplies obliged most countries to rethink their position vis-à-vis subsidies. But there are still several constraints affecting the review of subsidies.

Keywords Credibility · Subsidies · Regulation

Levy and Spiller (1994–1996) developed the concept of credibility around the ability of the regulatory system to protect private investments from opportunistic intervention of the legislative and executive bodies. The nature and extent to which each industry function is regulated varies. In MENA countries which have reformed their electricity sector, the activities of power procurement, transmission, distribution and supply of electricity are performed by statutory monopolists in their respective authorized areas. These activities are therefore subject to direct and extensive regulation. Regulation of the activity of generation is lighter. Generators have to compete to enter the market, and this determines the terms of the new entrant's remuneration, in the long-term Power Purchase Agreements (PPA). Properly structured PPAs should provide strong incentives for efficiency. These incentives, in conjunction with competition to enter the market, obviate in theory the need for extensive regulation of generation.

L. Benali, *Electricity-sector Reforms in the MENA Region*, Perspectives on Development in the Middle East and North Africa (MENA) Region,
https://doi.org/10.1007/978-3-319-96268-9_7

7.1 Credibility; the Role of Regulation, Performance Monitoring and Integrated Policies

As detailed earlier, the main need for regulation of generation is in relation to the share of the market which an individual investor is entitled to hold.[1] Market share restrictions are enforced either by law or in the licenses. Competition and auto-regulation are assumed to be able to deal with the rest (performance, delivery cost…). More generally, Spiller and others[2] have developed analytical frameworks where *the credibility of a reform is defined by its capacity to provide stability of commitments*, and the ability of its governance structure to solve regulatory problems between the government, the operators and the other stakeholders. The stability of commitments is a given if the three following and complementary conditions are fulfilled (Levy and Spiller 1994):

- Substantive constraints on the discretion of the regulator
- Formal or informal constraints in changing the regulatory system
- Institutions that enforce the above constraints

As described in the section on attractiveness and feasibility, the concept of credibility is linked to the institutional arrangements, mainly to their efficiency. The research of Holburn and Spiller (2002) and that of Perez (2002) highlight an additional need for *flexibility* of the governance and incentivizing structures. It was highlighted earlier how a rigid, particularly if decentralized, structure could affect the feasibility of the reforms. This point on flexibility is critical particularly in non-mature markets of developing countries. It conflicts with a debatable US-led political trend which states that strict separation of powers (like in the American institutional model) is a desirable model for stability of engagements. The credo is *« the more veto gates, the more difficult policy is to change, but therefore the more stable and credible it tends to be »*.[3]

Moreover, it appears important to refer to Wallsten (2001a, b) work on deregulation of telecommunications which concludes that there is no evidence of a significant influence of the existence of an independent regulator on reforms' success. However, a critical dimension of the credibility of the reform lies in the design of "effective" regulatory institutions. There is a large array of literature that defines the meaning of an effective regulatory system. Stern and Holder (1999: 43), for instance, suggest that regulatory institutions should be characterized by *Clarity* of roles and

[1]Economic theory and scrutiny of market behavior suggests a positive correlation between market share and market power, although the degree and nature of correlation is sensitive to market structure.

[2]Levy and Spiller (1994–1996) reforms of Telecommunications sector, Guasch and Spiller (1999) reforms of network industries, Savedoff and Spiller (1999) reforms of water distribution, Holburn and Spiller (2002) on electricity reform.

[3]Tsebelis (1995), but also McCubbins et al. (1987, 1989), Levy and Spiller (1994, 1996), and Haggard (2000).

objectives, *Autonomy* from political intervention, wide *Participation* by (or consultation with) relevant stakeholders, *Accountability* to outside agencies, *Transparency* of decision making process and *Predictability* of decisions.

As regional experiences show, each of three criteria is usually fulfilled: clarity of roles (as decrees and laws are usually inspired from global reform leaders), participation of relevant stakeholders (maybe too much, particularly in societies built on consensus) and accountability (to the highest authorities).

The transparency of the decision-making process is often guaranteed by economic regulation and open bidding procedures, except under specific circumstances (lack of bidders, intervention of stakeholders pursuing "national" interest). The autonomy from political intervention and the predictability of decisions are usually less certain. For example, it appears difficult to implement time-varying tariffs even if several regulators, or relevant decision-makers, are convinced of the benefits. One way would be to link them to other state policy goals, such as energy efficiency, energy independence and environmental improvements. In any case, regulatory authorities have to "implement the general policy of the State in relation to the electricity and related water sector (in Oman), including obligations arising from international agreements (provided such obligations are not in conflict with the Sector Law".[4]

One important dimension which applies to MENA countries is the shifting paradigms of economic regulation in other parts of the world.[5] Electricity reform and economic regulation methods in MENA have followed international trends. The timing and the sequencing of these reforms is important in that regard (Wallsten 2002). As reform vanguards and economic regulation leaders are reviewing the traditional models, in view of their maturity, of the reality of changing circumstances and of the rise of new challenges (climate change), so are MENA countries. *The main implication is a higher involvement of public authorities to ensure that the challenges inherent to their own countries (changing fuel supply costs, subsidies and sector's efficiency, climate change debate, local pollution) are taken into account.*

Another measurement of regulatory enforcement would be access to data and information. Market players could derive major advantages from asymmetries of information. Even when it is not done on purpose (rudimentary markets), just the mere fact that data is not easily accessible, or is not centrally deposited, indicates that regulation and market oversight is not enough enforced. The development of new technologies has facilitated this part of the evaluation exercise. Even if economic regulation's major weakness lies in its assumption that regulated entities respond to economic incentives, such regulation enables to highlight performance deficiencies. *Transparency and accountability could be improved simply by reporting and benchmarking performance. Therefore, the data accessibility aspect is critical.*

---

[4]Article 25 of Oman Electricity Sector Law, detailing the principal functions of the Regulatory Authority.

[5]See M. Pollitt « the future of electricity and gas regulation in a low-carbon policy world », Energy Journal, 2008 special issue on the future of electricity (papers in honor of David Newberry).

More concretely, the design of effective regulatory institutions hinges on the availability and the use of an adequate supply of economic resources, particularly the supply of trained staff, both in terms of number and in terms of adequate human capabilities (lawyers, technicians, economists…). A strong professionally trained administration should also be better able to maintain policy stability in the face of political instabilities. Domah et al. (2002) modelled the Costs of Electricity Regulation as related to human resources and showed evidence of Human Resource Constraints in Developing Countries. They conclude that in many large developed countries, such as the UK and the USA, regulators do possess most if not all of the "desirable characteristics" owing to their favourable endowment of human capital and legal institutions, but the economic and social structure of small developing countries may act as real obstacles in achieving them. Large developing, middle income nations such as Argentina, Brazil, Chile, Malaysia and the Philippines appear to have been able to cope with the staffing needs of their regulatory institutions, but for low income Asian and African countries and most of the small island states (such as Grenada or St. Lucia), the problem of the supply of adequate human capability is much more acute.

Most gulf countries have been able to attract high calibre economists from western countries, Asia or the rest of the Arab world to prepare, design, implement the reforms and then to follow up in the new institutions. The availability of appropriate and qualified human resources makes the regulation more "efficient". According to the Omani Authority for Electricity Regulation, the cost of regulating the electricity and related water sector in 2007 compared favourably with international benchmarks. It was less than one fifth of 1% of total sector turnover.[6]

Other countries in the region, particularly in North Africa, struggled to attract the few human resources they had in order to build up and maintain the institutions, largely because of the "brain drain" they were suffering from in most sectors of the economy. Table 7.1 offers a comparison between selected regulatory agencies in MENA and three of their European counterparts.

The development of robust regulatory frameworks and strong institutions can be hampered by under-funding and a reluctance of authorities to transfer independent decision-making to regulatory authorities (even when required to do so by Law). Since many regulators will begin performing their functions with little independence and capacity, transitional arrangements are needed (initial price controls in contracts or licenses or clear tariff setting principles in legislation). Contracting out some of the technical functions of regulation could also be considered.

Enhancing regulation should be a continuous and evolving process. The reality is that this is easier in some countries than in others. Abu Dhabi could modify its Law n°2 of 1998 concerning the water and electricity sectors in 2007 (Law n°19). The amendment expanded the range of regulated activities (from 12 to 16) and included for the first time the bulk supply of electricity. More importantly, the reporting

[6]Or around 2 Omani Rials per customer account, or a tenth of a Baiza per kWh supplied.

Table 7.1 A comparative evaluation of selected MENA regulators with CRE (France, OFGEM (UK) and CER (Ireland)

	Appointment(s)[a]	Reporting	Funding	Staff
OFGEM (UK)[b]	Open competition for chief executive	Gas and Elec Markets Authority (Exec. and non-exec. Members)	License fees	295 (2008)
CRE (France)[b]	President of the Republic, Parliament, Senate, Consumers	Independent (auditioned by the Assembly and Senate)	State Budget	129 (2008)
CER (Ireland)[b]	Minister of Communications, Energy and Natural Resources, Minister of Finance	Minister of Communications, Energy and Natural Resources	Levy on energy undertakings activities	59 (2007)
ECRA (Saudi Arabia)	Royal Order	Board of directors chaired by Minister of Water and Elec., Governor of the Authority as deputy chair, 6 senior government officials (min. of Water and Elec., Finance, Petroleum, Commerce and Ind., Economy and Planning, and SWCC); 5 members selected on own merits	License fees and fines	119 (2008)
AER (Oman)	Council of Ministers	Financial/administrative autonomy, but not functionally independent, final tariff approvals by Ministry of Finance	License fees	19 (2006)
RSB (Abu Dhabi)	Chairman of the Executive Council	The Executive Council of the Emirate	License fees	40 (2008)
CREG (Algeria)[b]	Prime Minister	Independent (but logistically and financially dependent on the Ministry of Energy and Mines, as the latter approves levy)	Levy on electricity/gas sales (0.07 cDA/Th to 0.75 cDA/kWh)	49 (2008)
ERC (Jordan)	Council of Ministers (recommendation of the Prime Minister)	Financial/administrative independence but "related to the Prime Minister" (Law)	License ans service fees. Grants and subsidies. (surpluses accrue to State Treasury)	86 (2007)

(continued)

Table 7.1 (continued)

	Appointment(s)[a]	Reporting	Funding	Staff
EERA (Egypt)	Prime Minister	Prime Minister	Licence fees and service proceeds. Funds from the State budget (not used)	50 (2008)

Note: This analysis could be completed with details on the terms (duration and renewal) of commissioners and key staff
Source: Regulators and Sector laws
[a]Appointment of key positions
[b]Gas and Electricity Regulation

structure of the Regulation and Supervision Bureau was changed from ADWEA to the Executive Council.[7]

Some countries decided to engage in a reform process without the creation of a regulation authority. Examples include Morocco, Lebanon and Bahrain. The tasks of electricity regulation are either left to a ministry (Finance, Electricity or Energy, General Affairs of the Government), to the Competition authorities if any or to the contract between the private sector and the consumers' representatives ("corporatist" solution). This last option bears cost and risk advantages for the State, but in general, an effective regulation in the absence of a regulatory authority assumes that the controlling role of local collectivities and professional associations is reinforced. Moreover, in these cases, the regulation would be:

- Mainly non-discretionary,
- Usually ex-post (while a regulatory authority would regulate ex-ante),
- Less information-intensive
- Discontinuous
- Dependant vis-à-vis the political framework.

Ex-ante regulation enhances access to information on the sector, particularly the information that regulated entities would have preferred to conceal ex-post. It also enables more predictability and transparency in actions and allows setting up performance incentives.

Ex-post regulation allows for less pressure on the decision-making process and for sanctioning a behaviour that has *actually* occurred. However, *in a context of vertical integration (as in Morocco, and Lebanon for instance), network managers could easily tend to deploy opportunistic behaviours to impose high tariffs for third-party access to the network, hereby maximizing their profits on the activity of transmission* (Brunekreeft and Keller 2001). This enables them to minimize their

[7]The Executive Council was formed in 1974 to replace the Council of Ministers created in 1971. It is the local authority of the Emirate and assits the Ruler to carry out his duties. It is chaired by HH Sheikh Mohammed bin Zayed al Nahyan, Crown Prince of Abu Dhabi and Deputy Supreme Commander of the UAE Armed Forces. Its membership consists of chairmen of the local government departments, some local authorities and other members appointed by the Ruler.

margins on competitive activities (generation and retail) and to indirectly limit competition.

In any case, regional experiences show that the coexistence of a strong State and a dependent (sometimes weak or absent) regulator did not prevent the move to the single buyer model and the introduction of competition at the Generation level in these countries. It sometimes expedited the process. Stability of commitments has been "guaranteed" by sovereign guarantees, in theory to reassure project lenders and operators that the government will take or refrain from taking certain actions that affect the project. The government is in any case better able to manage exchange rate risks and political risks. Sometimes, when the value of the sovereign guarantee is constrained by the sovereign debt, other mechanisms are pursued like insurance from multilateral, bilateral or export credit agencies. Private insurance companies have rarely come into play in power projects in the region. Finon (2004) states that it makes sense, in developing countries, to "enforce commitments in investments by creating public governance by long-term coordination and reduction of investment-related risk through long-term contracts". In comparison with the vertically integrated state-owned monopoly, public authorities will be kept at a distance but not completely set aside, even when competition is introduced. Actually, the single buyer model tends to shift most of the risk to the single buyer. Therefore, public authorities should remain partly/indirectly involved to induce efficiency incentives and investments signals.

One impact of high oil prices has been accrued revenues to the hydrocarbon's exporting countries. Governments might be less willing in the future to give sovereign guarantees if competition for projects increases, and if some competitors can find other ways to mitigate the risks without transferring them entirely to the State. As a reminder, successful IPPs in the region were built on the credibility of the entities, on the predictability of decisions and on the clarity of the processes. However, even if the institutional credibility is improved after the introduction of competition, the economic risk of investments remains.

7.2 Another Aspect of Credibility: The Subsidies Question

The credibility of these reforms could not be analysed without looking at the subsidies issue and pricing of energy (electricity and fuel). In some MENA countries, electricity is affordable for the majority of citizens, even free for some nationals. In most, the electricity sector is subsidized by applying a "double subsidy". For example, generators in some gulf countries pay a below-market gas price of around \$1–1.5 per MMbtu and the government subsidizes distribution companies

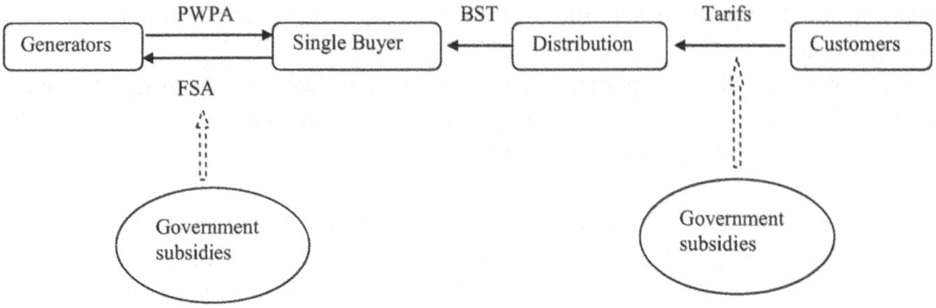

Fig. 7.1 The double subsidy and the financial transactions in the single buyer model

to cover the difference between the economic cost of electricity and actual tariffs to the end-users (Fig. 7.1).

When considering energy price reforms, governments did not initially place a high priority on considering the case for restructuring retail electricity prices. Electricity retail tariffs (or even the provision of the service for free for specific categories of the population—nationals vs. expatriates for example-) are often driven by social or developmental factors. Nonetheless, some like Dubai or Oman have already moved to cost reflective electricity tariffs as electricity subsidies growth could be significant with the increase in sales.[8] Saudi Arabia is also moving towards establishing cost-reflective electricity tariffs, while reforming fuel prices in parallel. That growth in the volume of subsidies is usually not offset by the hypothetical decrease in production costs due to efficiency and competition.

Managing the electricity subsidy would have to be through managing either demand (Demand Side Management programs) or supply costs or both. However, it appears that the very first step would be to increase transparency in electricity sector subsidies, e.g. by merging the "double subsidies" into a single item in governments' budgets.

Increasing gas prices would encourage generators to invest in energy efficient technologies, but would also increase the cost of electricity, hence increasing the level of direct subsidies to the electricity sector, assuming the same retail price. For a gas price of around $1 per MMbtu, regional wholesale prices were around 2–2.5 cents/kWh and for a gas price of $1.5 per MMBtu, experiences in the region show that wholesale prices would increase to 3–3.3 cents/kWh.

Simulation models were built to compare these wholesale prices with theoretical generation costs under different assumptions for projects locations and therefore for fuel costs and other hypothesis. The objective would be to forecast the actual and ideal generation costs under different fuel prices, assuming different levels of efficiencies of the generation fleet.

In a simplified version, total costs are represented as follows:

[8]Experiences elsewhere illustrate that if not dealt with properly, subsidies could skyrocket figures. By way of comparison, Egypt's energy subsidies increased from $220 million in 1999 to $7.7 billion in 2006.

$$\text{Total costs} = \text{capital charge} + \text{fixed and variable O\&M} + \text{fuel costs}$$

The simulation confirms that generation costs in the selected sample of countries and projects are indeed around the same order of magnitude. They vary between $16 and $23 per MWh (1.6–2.3 cents/kWh) when gas prices vary between $0.5 per MMBtu and $1.5 per MMBtu.

In the region, a mix of gas/fuel pricing for different sectors has developed over time, and the current use of subsidized energy prices has been successful in supporting the development of the electricity, industrial, and export sectors. This approach has historically been adopted by resource-rich nations in the region, for example Oman's gas prices have been at typically around $0.8–1.5 per MMbtu, Saudi Arabia has been selling its wholesale gas at $0.75 per MMbtu and in Egypt, several industrials used to pay $1.2/MMbtu while the Egyptian government purchased the gas from upstream producers at more than $2.65 per MMbtu. Iran provided the gas to the national power utility at the $0.1 per MMbtu, to industrials at $0.6 per MMbtu and to the residential/commercial sector at $0.45 per MMbtu (Table 7.2). World Bank 2006 stated that the explicit subsidy in Egypt is about $7 billion a year and estimates the same subsidy in Iran at $6.5 billion. By the end of the 2010s, several countries have been considering at least on paper to reform their gas prices, particularly as a series of long-term contracts were reaching expiry during the 2010–2020 period.

Energy subsidies are often difficult to estimate, and a subsidy delivered through a gas price has the disadvantage of clouding whether an industry or an electricity generator is competitive or value adding, as this may not be evident from company's financial statements. Moreover, a subsidy applied through fuel pricing can potentially be counter-productive to stimulating efficient use of energy. It is only when energy supplies are restricted or when the subsidy scheme might contravene some international regulations[9] that the limits of the approach become evident. In parallel,

[9]Saudi Arabia's potential subsidizing the petrochemical industry through low gas prices has been a major issue when the country was discussing its accession to the WTO. The report of the Working Party on the Accession of the Kingdom of Saudi Arabia to the WTO of November 2005 detailed the Saudi position and arguments in this regard. More specifically, "the representative of Saudi Arabia stated that producers/distributors of NGLs in Saudi Arabia would operate, within the relevant regulatory framework, on the basis of normal commercial considerations, based on the full recovery of costs and a reasonable profit. He confirmed that his Government's policy was to ensure that these economic operators, in respect of their supplies of NGLs to industrial users, would fully recover their production and investment costs and make a profit in the ordinary course of business." Also, "local NGLs sales have a higher relative commercial value to the producer than sales to seasonal customers in export markets that purchase NGLs for uses such as fuel, with seasonal swings in demand". On the other hand, "natural gas was not sold for export due to the high costs of liquefying, transporting and re-gasifying such gases, and therefore had no international reference price in the Gulf region. Previously, natural gas had been burnt as a waste product, but was later collected and made available to all interested users on a non-discriminatory basis, at a regulated price of 281 Halalas per million BTU (US$ 0.75 per million BTU) until 2015. This decision was taken based on a combination of commercial reasons and environmental concerns". Saudi Arabia became a full WTO member in December 2005.

Table 7.2 Generation costs under different assumptions for projects' location

Country and Energy type		Libya Gas 1	Saudi Arabia Gas	Qatar Gas	Libya Gas 2	Algeria Gas	Abu Dhabi Gas	Egypt Gas 1	Egypt Gas 2	Iceland Hydro	South Africa Coal	Iceland Geothermal
EPC cost	$/kW	600	600	600	600	600	600	600	600		1000	
Owners' cost	$/kW	200	200	200	200	200	200	200	200		250	
Capital cost	$/kW	885.0	876.2	875.6	885.0	867.7	877.2	885.0	885.0	2348	1250	2000
Interest rate	%	7%	7%	7%	7%	7%	7%	7%	7%	7%	7%	7%
Fixed O&M	$/kW-year	20	15	15	20	20	15	20	20	20	25	80
Variable O&M	$/MWh	2	2	2	2	2	2	2	2	0	3	0
Heat rate	Btu/kWh	7,082	7,074	7,073	7,082	7,066	7,074	7,082	7,082		10,000	
Fuel price	$/MMBtu	0.50	0.75	0.75	0.80	1.00	1.00	1.25	1.50		0.72	
Capacity	MW	1200	1200	1200	1200	1200	1200	1200	1200	1323	1300	200
Capacity factor	%	86%	87%	87%	86%	88%	87%	86%	86%	90%	90%	90%
Annual production	MWh	10,000,000	10,000,000	10,000,000	10,000,000	10,000,000	10,000,000	10,000,000	10,000,000	10,437,000	10,249,200	1,576,800
Life	years	30	30	30	30	30	30	30	30	50	40	30
Capital charge	$/MWh	8.56	8.47	8.47	8.56	8.39	8.48	8.56	8.56	21.57	11.89	20.44
Fixed O&M	$/MWh	2.40	1.80	1.80	2.40	2.40	1.80	2.40	2.40	2.54	3.17	10.15
Variable O&M	$/MWh	2.00	2.00	2.00	2.00	2.00	2.00	2.00	2.00	–	3.00	–
Fuel	$/MWh	3.54	5.31	5.30	5.67	7.07	7.07	8.85	10.62	–	7.20	–
Total		16.50	17.58	17.57	18.62	19.86	19.36	21.81	23.58	24.10	25.26	30.59

Project location		Tripoli	Dhahran	Doha	Tripoli	Skikda	Abu Dhabi	Cairo	Cairo	9 projects	Subcritical PC	Assumes flash steam
Max temperature	°C	48	47	47	48	44	47	48	48			
Elevation	meters	81	17	10	81	7	27	81	81			
Elevation effect	%	99.1	99.8	99.9	99.1	99.9	99.7	99.1	99.1			
Temp effect	%	91.2	91.5	91.5	91.2	92.3	91.5	91.2	91.2			
ISO capacity	MW	1327	1314	1313	1327	1302	1316	1327	1327			
Average annual temp.	°C	20	26	27	20	17	27	20	20			
Part load	%	90.4	91.3	91.4	90.4	92.2	91.2	90.4	90.4			
Relative heat rate		98.8	99.0	99.0	98.8	99.1	98.9	98.8	98.8			

Note: Weather & altitude data from Weatherbase.com

Fixed O&M were discounted by $5/MWh to reflect lower labour costs in gulf countries. This might need to be revised in the medium term, and is reflected in North African example an increase of 100% in gas price would increase generation costs by 45%.[10] But the mindset has been changing: in some countries where indigenous fuel supplies become scarce, being heavily committed to contracts signed during the 1990s (when the first IPPs were launched and when market prices were significantly lower than today) becomes problematic. Regionally, the value of natural gas is becoming increasingly linked to at least its cost of production or to its export netback value.

A few countries have started to negotiate more sophisticated terms (gas prices indexed to end-user pricing or including a sharing of end-product netbacks) in recent contracts with some industries. Higher prices are also being used for project screening (typically $3.00/MMBtu) in anticipation that high prices might prevail. This is already driving ideas to introduce more energy efficient technologies (e.g. CCGTs in the power sector instead of simple cycle), which could be otherwise imposed by law as well. Even in the upstream sector, the installation of energy efficient power generation systems has become an important factor in the planning of projects.

More generally, energy subsidies are the sum of subsidies injected in various parts of the demand chain: households (gas, electricity, refined products), industry (gas, electricity), Electricity (gas, oil and refined products).[11] In a simple framework, they can be calculated relative to market prices—opportunity cost- or to production costs. For example, a domestic gas price of $0.8–1.5 per MMbtu could be compared with an estimated medium term market rate of $4.5 per MMbtu, or with the domestic production cost.

Razavi (2009) was the first to provide calculations of Long-Run Average Incremental cost (LRAIC) as a function of the Long-Run Marginal Cost of gas supply, at the wellhead and city gate. Based on IEA 2005 forecasts of production and investment needs, he estimates that the LRAIC at the wellhead varies from $0.4 per MMbtu in Algeria and Saudi Arabia to $0.6 per MMbtu in Qatar. At the city gate, the minimum LRAIC in the region goes from $0.61 per MMBtu in Saudi Arabia to more than $1.1 per MMbtu in Iraq given the high transportation investment needs. He then adds the depletion premium to the LRAIC in order to calculate the economic cost of the gas at the city gate in each country.

Nevertheless, even if Razavi's analysis is interesting from a comparative perspective and helpful from a numerical perspective, it bears some limits. It is based on investment estimates which date back to 2005, during the early part of the capital costs escalation. Therefore this cost inflation has not been factored in his calculation. Moreover, it disregards the cost of the "marginal btu", which would be more relevant in tariff applied to new entrants as new projects come onstream to serve these new entrants. The cost of the marginal btu would reach $4 per MMBtu for offshore gas (Egypt) or $5 per MMbtu for sour gas (Abu Dhabi).

[10]We applied a gas price of $3 per MMbtu to the Egyptian gas fired utility.

[11]According the IEA, 20 non-OECD countries, including China, India, Russia, Iran and Saudi Arabia, spent $310 billion to subsidize oil, gas and power in 2007.

In parallel, the depletion premium is defined as "the opportunity cost of consuming a unit of the exhaustible resource now rather than in the future". It is estimated based on the cost of the substitute fuel (if any) at the time that the depletion constraint becomes binding and consumers would need to shift to another fuel. When they wanted to encourage utilities to shift to gas, MENA countries did not impose a depletion premium on oil but rather offered a low gas price. Razavi offers a helpful depiction of the concept of depletion premium and calculates that this premium varies from $0.09 to $1.31 per MMbtu in 2005 for example for Iran and Algeria respectively. Based on the current knowledge of reserves, the depletion premium increases over time and reaches $0.71 per MMbtu for Iran and $8.83 per MMbtu for Algeria in 2030. Iran continues to have the minimum depletion rate among the selected countries and Algeria will be taken over by Libya ($9.21 per MMbtu in 2030). According to his analysis, Iran has the most distant switch date (2057).[12]

Project financiers are aware of the limits of the depletion rate concept which they usually mitigate by operating a number of sensitivity analyses. Some of the main uncertainties relate to the changing knowledge of the size and life of the deposits, the actual substitute fuels and the future prices of the gas and of its substitutes. But one of the main limits is that this analysis falls short of properly addressing the case of countries where reserves are available but not easily deliverable, and where the production level is not driven by reserves size, but by above-ground factors. All countries have their hydrocarbons production limited to some extent by other factors than the geological ones. However, some are unable to make their productive capacity rise to a level anywhere close to the technical limits of their reserves potential. The best example in the region is Iran.

Indeed, Iran holds the world's second largest gas reserves. Yet, its gas sector has been gripped by a series of acute crises, creating domestic and regional discontent. Iran also has difficulty bringing major export projects to fruition as it faces a set of specific issues related to the institutional set-up in the country, the infrastructure limits, the lack of investments and the strained international relations.[13] Complex organization of Iranian politics and the energy sector slows the decision-making process. Infrastructure is insufficient and creates bottlenecks. Major investments are needed on all fronts, but international pressure and sanctions discourage potential operators and hampers the financing of projects. The country will struggle to meet its own needs; strategic "choices" between domestic demand and exports could become irrelevant given demand pressures. Iran confronts tough strategic choices regarding its industrial policy and management of its domestic demand and supplies. Gas consumption will continue to increase by more than 8% annually, driven by residential and industrial demand, but also by substantial needs for reinjection in oil fields. Gas balance will remain tight, with further shortages in the upcoming winters.

---

[12]The date of switching from gas is estimated by extrapolating demand to the point at which a production plateau of 15 years can be supported based on proven reserves.

[13]See also Chevalier (2004), *les Grandes Batailles de l'Energie*, pp. 337–338.

Domestic needs will take precedence over existing exports, and the upstream oil sector will compete with the power and industrial sectors for gas supplies.

In this context, the depletion premium is distorted by levels of production which do not reflect the country's potential and are not a result of a willingness to delay the production of a unit of gas by some years. It appears indeed difficult to talk about "choices", "decisions" and "opportunity costs" in the context of the Iranian gas conundrum. The Iranian authorities are obliged to delay the production, even if local demand is pressing and exceeding actual productive capacity. Going forward, in order to assess costs, opportunity costs will be used only when domestic demand is all served and when opportunities actually exist (possibility to consume now rather than in the future, and vice versa; possibility to export with export outlets and markets).

It appears important to note that subsidies are naturally highly sensitive to international oil and gas prices and have therefore grown considerably since 2000. They are also dependant on the size of the specific market, so for example, the fuel subsidies to the industrial sector increase in countries which pursued an aggressive industrialization policy, in view of the rapid growth in the sector. In all these countries, energy demand is supported by several factors, including subsidized domestic retail tariffs but also other "structural" aspects like high economic growth, post-war reconstruction, lack of public transportation, large young populations, and industrialization. Therefore, the real impact of subsidies on demand is difficult to measure, as is price elasticity of demand. In most cases, it is even difficult to quantify the actual amount of energy subsidies being injected in the economy. Egypt is one of the rare countries in the region to publish the details of "social spending" in its annual budget. In fiscal year 2008–2009 (starting July 1), subsidies for petroleum products were expected to reach \$11.7 billion, up from \$1.1 billion for all subsidies in 2002–2003. The share in the national budget (\$52 billion today) also increased from less than 6% to 22% in the same timeframe. In most other countries, governments have yet to have a clearer idea on the actual subsidies' bill. However, a clearer view would be important before attempting to introduce demand measures.

Energy subsidies are a substantial part of the country's budget and they continue to grow if the situation remains unchanged. *It is important to develop a clear view of their size, make them transparent and transfer them into a single "subsidies pool" for example. It is also important to consider electricity subsidies in a comprehensive analysis that includes all energy subsidies and not in isolation, as governments usually look at the total energy subsidies when considering tariff reforms.*

Barring a few exceptions, and until 2014 when the second waves of energy price reforms in the region occurred, end-users in the region have generally been insulated from high international energy prices due to highly subsidized environments. However, the combination of high oil and gas export prices, with the temptation to free up more hydrocarbons for exports, and of exploding demand, hereby increasing subsidies needs, introduced a new level of urgency and spurred some to effectively introduce measures to rein in demand. Recent events show an important evolution with the transition to international market prices in regional energy transactions.[14] Gas exports deals are still oil-linked (long term large volume LNG contracts from Qatar to China) and regional gas pipeline prices are much higher than the traditional $1 per MMbtu. Dolphin's main long-term gas supply agreements (GSAs) contain escalation factors, and the short-term, interruptible GSAs (Ras al Khaimah, Dubai, Sharjah) are priced at levels considerably higher than those for long-term contracts. Negotiations between Oman and Iran signal that the only way to secure a cost-plus price is through equity in the development of upstream projects, when possible. Even then, the ultimate utilization of the gas (domestic or for exports) is taken into account in the price negotiations. Last but not least, Dubai, despite being one emirate of an LNG exporting country, was set to be the first LNG importer in the region with the $450 million floating storage and regasification scheme starts delivering gas in 2010. Actually, Kuwait started importing LNG in the summer of 2009. In other words, Middle Eastern countries are seeing the price of their gas supplies (indigenous or imported) increasing.

Therefore, some governments stepped in to introduce measures to manage local demand. "Soft" approaches are already taking place in various countries with mixed results.[15] Countries where the subsidies' bill becomes a real burden on the government's budget or which have to rationalize energy consumption because of scarcity of fuel tend to be the first ones to introduce "harder" approaches:

"Soft" approaches: Governments are responding to tight local and global markets by tackling inefficiencies, wherever possible. This includes efficiency/technology requirements particularly on new projects, demand-side-management measures and energy efficiency promotion campaigns. Even if energy is becoming more expensive, several governments would still be reluctant to adopt "harder" approaches like passing through costs or export price netback to their local populations.

"Hard" approaches: Most countries are rethinking the energy subsidies' rationale. A first wave effectively introduced tariff reforms or rationing schemes before 2013. Usually, the countries where these measures were introduced are the ones which are the most hardly hit by high energy prices or other constraints (such as vulnerability because of international sanctions, declining domestic production).

- Syria, which spent $8.7 billion in 2007 on diesel subsidies, introduced in April 2008 a diesel rationing scheme, allocating a quota of 1000 L/month per family, and had already increased for the second time in a few months retail gasoline

[14]For example, about 66% of contracted Qatari LNG goes to Asian customers, on long-term oil-linked contracts using the so-called S-curve to protect exporters against low oil prices and consumers against high prices. An additional 16% is contracted to Europe.

[15]"Price is generally the go-to solution if you want to do something quickly (about the rising demand for water and electricity). But if you are looking at affecting the curve over the long run, and not necessarily having to install all that capacity—3, 5, 7 years out in the future—then you need to educate and apply a softer approach." Erik Jankel, chief executive officer of Norway's Aqualyng, 9 December 2008, MEED.

tariffs (87 cents/L). But at the same time, the viability of these measures is usually questioned in the absence of a comprehensive economic program. In Syria, it is still unclear how and when the promised package of essential economic and social measures to counterbalance retail prices' increase (salary increase, health insurance) will be introduced.

- In Jordan, refined products' prices are set on a monthly basis and subsidies have been lifted in 2007 on all, except on LPG.
- Iranian authorities had to introduce some changes to the gasoline rationing scheme introduced in June 2007: private drivers' quota was increased from 100 to 120 L/month (26–31 gallons) at the same subsidized retail price of 1000 rials/L (34 cents per gallon). Drivers are allowed now to buy above that quota, at a higher but still partly subsidized price (4000 rials/L). The most recent change (June 2008) introduces a price difference depending on the type of cars. The new scheme makes one step closer to market prices.[16] Iran attempts to gradually introduce the transition to higher retail tariffs while guaranteeing a vital minimum at subsidized prices. The Iranian rationing scheme did not reduce gasoline demand but contributed to decreasing demand growth (12% in 2006, 5% in 2008). The original aim was to cut gasoline consumption by 40%—to end Iran's dependence on gasoline imports in order to reduce its vulnerability to international pressure—.
- The Egyptian authorities have already taken several measures aimed at reducing energy subsidies.[17] In September 2007, they doubled domestic fuel oil prices to LE 1000 per ton (188$/t). In May 2008, they went further by increasing diesel prices by 60% and 90, 92-octane gasoline by 30%. 95-octane gasoline price nearly tripled. The new energy policy package includes also proposals to gradually revise upward the price of natural gas for export-oriented industries.

Globally, reforms usually bring concerns about inflationary pressures (first or worse second round effect), adverse social impact and loss of competitiveness. Where possible, governments responded by a range of targeted tools to mitigate such inherent concerns (cash transfers, vouchers, support on key needs, tax incentives, segregation of industries). A comprehensive plan is required to correctly diagnose and deploy appropriate tools. A successful reform case is defined as a case where (a) authorities did not have to reverse the reform and (b) it achieved a meaningful impact relative to its original objectives. Moreover, during multiple reform episodes in a given country, success could be achieved in one area of the

[16]It excludes super gasoline from the rationing system (priced at $0.59/L up from $0.15/L previously).

[17]Iran and Egypt have been developing a dual-fuel vehicle fleet, not only to tackle local pollution problems, but also to partially solve the gasoline issue. In Iran, the current number of CNG cars is estimated at around 50,000 to 70,000 cars out of a total fleet of 6 million, consuming around 1 BCM of gas. The authorities stated that they were investing $1.5 billion in four years to convert 1.5 million vehicles by 2010. However, these plans look overambitious in view of the potential gas constraints facing the country, the shortage of CNG cylinders and the concentration of the production in two local car manufacturers (Iran Khodro and Saipa), creating additional bottlenecks.

reform and not another. A reform journey/process may entail a 'pause' to reconfigure the implementation strategies in response to emerging conditions that were not original envisioned. Pricing reforms may also entail introduction of new subsidies that are deemed more targeted to the objective, but ultimately support the overall reform agenda, these often take the form of mitigation measures.

It is worth noting that, in general, successful reforms took over 10 years to plan, communicate and implement on a gradual basis to give consumers time to adjust. Whereas several countries were forced to reform due to fiscal pressures, in the most successful cases, a group of countries, including Turkey, China, Brazil, and Malaysia were proactive and initiated reforms when in a position of financial strength and/or economic growth. In addition reform in the energy sector should be part of broader economic reform to allow economic activity that gives households and industry an opportunity to sustainably absorb the price increases.

Increasing oil and gas market prices can act as an impediment to align domestic prices with international benchmarks. However, some countries with well-designed reform plans were able to successfully execute reforms (e.g. Turkey). Following the recent sharp reduction in oil prices, several countries, with reforms on the drawing board, are now accelerating their plans (Malaysia, Indonesia, GCC). By making retail prices market-reflective at a time of lower benchmark prices they are seeking to minimize the impact on their consumers' out of pocket expenses and reduce any adverse economic and social impact of reforms. They do not all necessarily aim at a full liberalization of retail prices. Some still opt for a managed float system (Malaysia); others keep an ad-hoc pricing system (Indonesia).

Countries have followed a range of different approaches. Each country's path is unique and subject to the specific prevailing global and domestic conditions coinciding with the timing of the reform. In some cases, reforms had to be rolled back or suspended only to be brought back or adjusted when circumstances were more favorable. In other cases, reforms were simply frozen. International best practices indicate that comprehensive reform architecture need to encompass five key elements:

1. A Comprehensive energy sector reform plan with accompanying institutional reforms
2. A communication strategy
3. Energy efficiency programs in parallel
4. Phased and gradual price increase if any
5. Targeted mitigation measures for vulnerable populations and economic activities

Even in cases where the authorities' commitment is strong, there are a few additional factors that might prevent or delay the lifting of subsidies. In most cases, governments face the insurmountable dilemma of the need to reduce energy subsidies without fuelling an already high inflation rate, and creating social

Table 7.3 Inflation clusters in MENA

Cluster 1: Mild-to-moderate inflation (rate $<8\%$)	Cluster 2: High inflation (rate $\geq 8\%$ and $<15\%$)	Cluster 3: Run-away inflation (rate $\geq 15\%$)
Slow pace: Morocco, Sudan,	Slow pace: Yemen.	Slow pace: Iran
Tunisia, Syria, Mauritania.	Medium pace: Jordan.	
Medium pace: Lebanon, Bahrain, Algeria.	High pace: Egypt, UAE, Kuwait, Saudi Arabia, Libya, Oman, Qatar.	

discontent. The flip side of the coin in relation to the recent high economic cycle has been inflation: high rates witnessed all over the region are linked to local markets' booming and increases in global food and imported goods prices.

Aissaoui (2008) defines three main clusters for inflation in the MENA region, according to the pace of inflation and its actual rate (Table 7.3).

Countries with inflation of more than 8% have one feature in common: they all have directly or indirectly (the case of Jordan) benefitted from exceptionally high petroleum revenues.[18] As a result, and despite having set budget resources significantly below prevailing oil prices to match their absorption capacity, both governments and the private sector have enjoyed unprecedented levels of liquidity supported by strong growth in money supply and relatively low interest rates.

In gulf countries more specifically, Inflation is exacerbated by the continued depreciation of the dollar (all gulf countries except Kuwait have their currencies pegged to the US dollar). De-pegging the currencies would have several drawbacks: Gulf countries continue to have significant assets in the US and the de-pegging would decrease the value in local currency of the hydrocarbons' exports. The decision would be a difficult one to make anyway given the long-standing alliance with Washington. As a consequence, it is highly likely that most Gulf countries would remain committed to the dollar peg in the short to medium term.

The second aspect which makes difficult the lifting of energy subsidies is the differences in retail tariffs between neighbouring countries. These differences create the conditions for the development of black markets and smuggling, but this is mostly relevant for easily transportable crude and refined products. This distorts the consumption patterns and might prevent any implemented demand measure to reach tangible results. In the case of electricity, regional markets are not fully interconnected and synchronized yet as detailed in Chap. 1. Assuming the regulatory framework would enable electricity trade, potential eligible customers are not able yet to technically purchase electricity in other countries in order to leverage tariff differences.

It is important to note the existence of cross-subsidies in some countries (see Fig. 7.2). Qatar and Morocco were examples where tariffs to the industrial sector

[18]The choice of this threshold stems from empirical evidence that, even though inflation is felt across the social spectrum at moderate level, it is more costly in terms of impact on GDP growth when it surges above 8%.

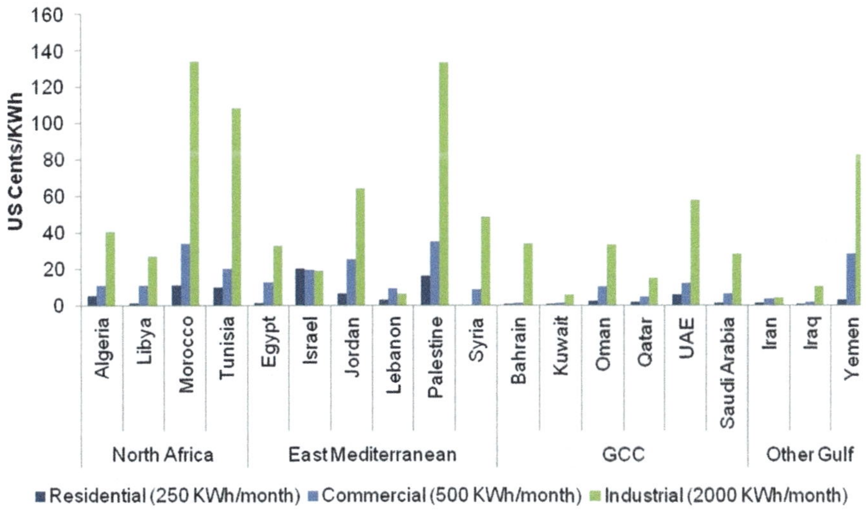

Fig. 7.2 Electricity tariffs in selected countries, 2012. Notes: Prices are averaged between (where applicable) dynamic night- and day-tariffs. Iran's highly fluctuating exchange rate makes a meaningful valuation of electricity prices in $ terms difficult. Exchange rate of IR30,600:$1. Iranian electricity prices are as of 2013. UAE prices are based on Abu Dhabi. In some cases, industrial users may also cross-subsidize residential and commercial users. There is no publically available information on such practice other than case-based narratives

have been lower than tariffs to households. The situation is the opposite in most countries in the region. These tariffs will be reformed later in selected countries but not in all.

As explained in Chap. 1 and in the "attractiveness" section, the desired industrialization contributed heavily to the growth in electricity demand in the region, and at times to the launch of reforms. Conversely, domestic fuel prices could be sometimes the only attractive advantage that the region can offer for energy intensive industries, even if, in some cases, such approach turned out to be counterproductive as far as efficiency and fuel supply are concerned. The following figure shows the competitiveness of possible aluminium smelters in Saudi Arabia using fuel oil or natural gas as fuels. Today's costs are obviously higher and this coincides with a downturn in the commodities markets, including aluminium. This means that projects are less urgent. As an example, one project, Ras al Zour, closed at an electricity price of $33 per MWh, much higher than the "conventional" $20 per MWh that made Middle Eastern smelters very competitive in comparison to their peers in other parts of the world (Fig. 7.3).[19]

[19]The $5.5 billion Ras al Zour project faced difficulties. Originally conceived as an IWPP, the project was turned into an EPC contract in April 2009, following a proposal of the finance ministry. Earlier in the month, the preferred bidder consortium, led by Japan's Sumitomo Corporation, fell apart after Malaysia's Malakoff pulled out, as it was not able to fulfill all its equity contribution, even if partners proposed to raise their stakes. Malakoff was originally due to take 20% in the project. The power project was already able to secure $2.5 billion in funding from the Japan Bank

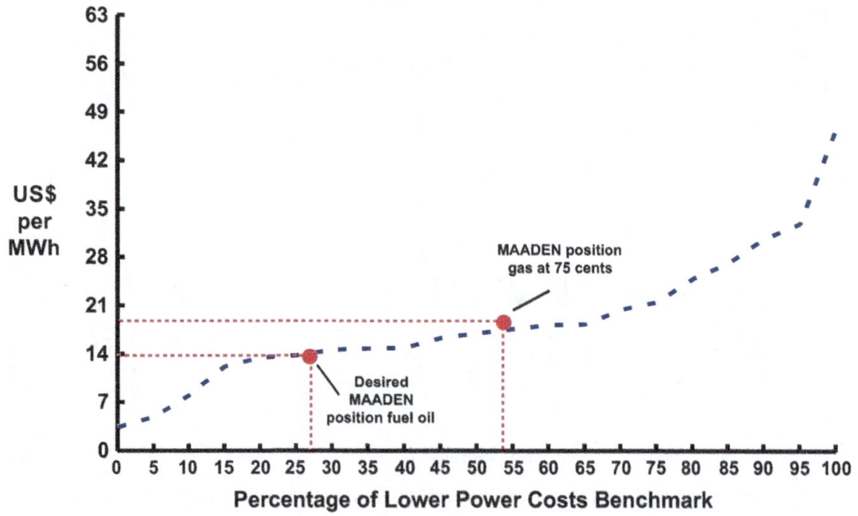

Fig. 7.3 Saudi Arabia's Maaden project competitive position globally. Source: Primary Aluminium Smelter League, Saudi Arabian Mining Company

As mentioned earlier, low pricing of fuel can stimulate international trade issues for products of energy intensive industries. For example, a dispute between Saudi Arabia and China emerged during the spring of 2009, about trade of only $2 billion per year, but bore highly symbolic issues, as it could apply to all energy-based products exported from the Gulf region. China accused Saudi Arabia of dumping Methanol into the Chinese market at prices below costs and threatened tariff retaliations. The latter would go against WTO rules, but the debate would take years if it were to be taken to the WTO. Methanol's producers in Saudi Arabia used to buy methane at $0.75 per MMbtu, 10–15% of industrial gas price in China. But the below-cost argument might not fly as most of Saudi methane is associated, and associated to one of the world's cheapest oil production.

In a period of economic recession, the issue is usually raised more vehemently as, in such downturn and low commodities' prices, low-cost manufacturers can survive while high-cost ones might have to shut down operations momentarily or permanently. This would cause more energy-intensive industries—and the associated jobs—to migrate to low energy costs places, until local governments decide to slow down or to reverse industrialization policies. For example, the European Union has imposed a 13% anti-dumping tax on Algeria's partly state-owned Fertial,

for International Cooperation. It was designed to supply electricity for an aluminum smelter and a fertilizer plant to be built in the new industrial complex north of Jubail. The power plant was to have a capacity of 1100 MW and 1 million cubic meters of water per day, and would be fueled with heavy crude.

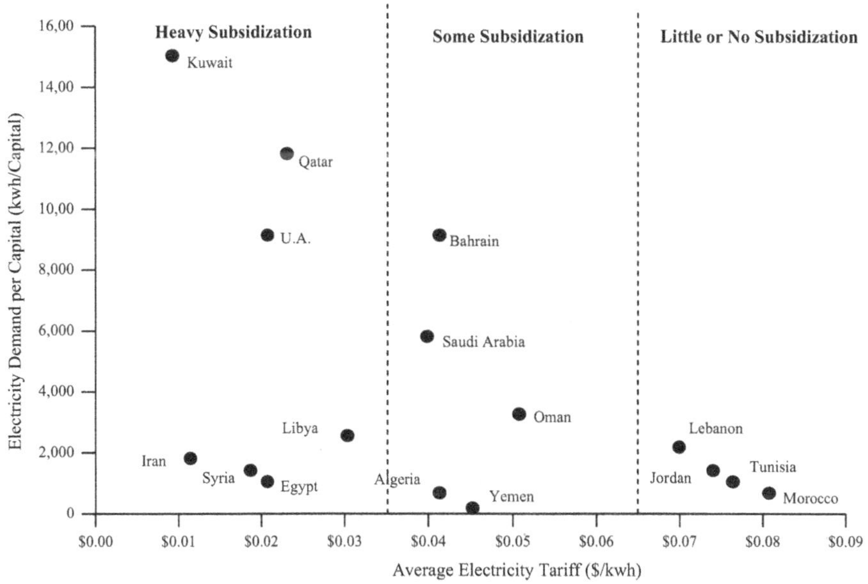

Fig. 7.4 Electricity Tariffs vs. Demand Per Capita (2010). Source: Energy information Administration, World Energy Outlook

while Algeria maintains that the internal gas price is real and does not constitute a subsidy.[20]

Finally, in the case of energy exporters, domestic tariff reforms are sometimes difficult to justify internally, and they usually question the overall pricing policy of fuel exports, or the mere export policy of the country (Fig. 7.4).

Iran, Oman, the UAE, Egypt have all experienced during the last few years a questioning of their export policy when authorities have raised the prospect of tariff reforms or of fuel supply tightness. Domestic (residential) needs usually take precedence over exports, however, hydrocarbons' resources—gas more than oil—need a market to develop. And countries who would want to monetize their natural gas resources quickly find themselves locked in long-term fuel supply agreements to industries and/or long term exports agreements. As the domestic market develops in parallel, availability of supplies (even in countries with large reserves) can quickly become an issue. Here again, the planning function is critical, but the challenge is that it should encompass all sectors of the economy (urban planning, industry, exports, upstream…etc) and bring together various entities (institutions, ministries, companies). Empirical experiences show that this becomes a great challenge in various hydrocarbons-exporting countries, and the task becomes so daunting that the

[20]Algeria's talks for accession to the WTO have been going for more than 10 years.

question of domestic tariff reform takes a back seat in the agenda.[21] At the retail level, it is usually better if the subsidies issue is completely delinked from the question of IPPs introduction.

More generally, there are other ways to assist the segments of the population which will still find the price an extreme financial hardship, even when electricity is affordable for the majority. To address this issue, such customers need to be identified and a payment mechanism established that makes it possible for them to have electricity. The most common mechanism is bill assistance.[22] In liberalizing markets, other mechanisms include:

- **General taxes**: The power industry is bypassed and subsidies are paid to low-income groups through the country's welfare system. It becomes a general redistribution policy within the tax system, unrelated to the liberalization of the electricity market.
- **System charges:** Public service objectives are funded by establishing a system-wide levy that is applied to the regulated distribution of electricity. The scope and application of system charges range widely among countries.
- **Special budget allocations:** Governments can also draw on public budgets (tax the general public). Alternatively, funds could come from external sources such as aid program for power sector related social programs and institutional reform.

A good reform is a feasible reform. It appears that there are mainly three steps needed in order to deal with energy subsides, if overnight tariff changes have to be ruled out for all the reasons detailed above:

The first step would be to merge the subsidies to the electricity sector into one line item in the government's budget. This will make the subsidies clear, visible, quantifiable and more manageable. The authorities could decide then whether they would continue to directly subsidize the sector or to replace these subsidies with other forms of subsidies. Moreover, fuel will be provided at its true cost of purchasing. In countries where no private upstream players are operating, this will have to be at a cost that would enable the National Oil Company to be financially viable.

The second step is to consider what would be an "acceptable" level of demand and demand growth (defined for example as natural growth resulting from additional households and additional economic activity) that would not hurt economic growth and minimum levels of access to energy. Anything higher than that level would be overconsumption and could be charged at full commercial cost (i.e. no subsidies). The latter could even include the coverage of future capital investment needed to cover this overconsumption. Equally, peak and off-peak pricing (either traditional or

[21]Based on author's observations in various projects advising governments in the region on their energy policy.

[22]These are recurring price subsidies that make electricity more affordable for low-income and vulnerable consumers. Bill assistance programs include discount programs (block or lifeline tariffs), percentage of income payment plans and arrearage management programs. They may also include emergency assistance to forestall service disconnection or ease the burden of short-term energy price spikes or extreme weather.

dynamic) would be introduced in countries where that is not the case yet (see Table 7.4). The approach detailed in the second step will come at no cost for the government. It could be defined as a passive approach. Countries where demand growth is reaching unsustainable levels should pursue more aggressively more proactive demand-side management techniques.

The third step would be to ensure financial viability of the electricity sector. Fuel would be delivered at cost and tariffs still subsidized (with the burden of these subsidies transferred from the utilities to the government, in countries where that's not the case). However, because of unpaid consumption (in addition to non-technical losses), the electricity sector will still not be able to break even. In countries where distribution companies were privatized, this became less of an issue since the private sector is generally better able to bill consumption, collect payments and deal with payment issues. But in most countries in the region, electric utilities are in a bad financial situation. Some even bear part of the subsidy burden when the government does not fulfil its subsidies' obligation on time. In these cases, the liability bore by utilities could be high, particularly when subsidies account for more than half of the price (see Table 7.5).

Moreover, the overall taxation policy of the country will need to be considered. Qatar, for example, which carries the double financial burden of inexistent of taxation and free tariffs for citizens, could be considered as an anomaly. In general, it is assumed that if electricity is provided by the government and not by the private sector, it could be financed by taxation (and provided for free as a pure public good, but recognizing the associated risks on demand distortions). Alternatively, subsidies could be financed by taxation, if electricity is provided at below-cost prices, since the government-owned utility would not be lucrative and would primarily seek to maximize social welfare.

Table 7.4 Retail tarrif setting in MENA

	Pricing per cons. tranche	Peak/off peak pricing	Lowest tariff (category) ($ cent/ kWh)	Highest tariff (category) ($ cent/ kWh)	Privatizat.-distrib.
Morocco	Yes	Not for low voltage	12.33 (residential <100 kWh/month)	19.71 (residential >500 kWh/month)	Partial
Algeria	No	Day/ Night	3.6 (Industrial)	5.7 (residential/ commercial)	No
Tunisia	Yes	Yes	6.34 (commerce<50 kWh/month)	14.91 (resid > 300 kWh/month and air heating and conditioning)	No
Libya	No	No	2 (residential)	5.2 (commercial)	N.A.
Egypt	Yes	No	0.93 (residential <50 kWh/month)[a]	9.6/10.2 (commercial > 600/1000 kWh/month)	Considered

(continued)

Table 7.4 (continued)

	Pricing per cons. tranche	Peak/off peak pricing	Lowest tariff (category) ($ cent/ kWh)	Highest tariff (category) ($ cent/ kWh)	Privatizat.-distrib.
Israel	Yes	Yes (time, day, season)	8.5 (street lighting)	12.2 (residential)	No
Palestine	No	Seasonal	9.89 (public lightening, Jerusalem)	14.3 (commercial)	N.A.
Jordan	Yes	Day/: night but only for SMI	4.5 (residential <160 kWh/month)	15.93 (residential >500 kWh/month) and 12..13 for commercial	2008
Lebanon	Yes	No	2.33 (residential and commerce <100 kWh)	13.31 (residential and commerce >500 kWh)	N.A.
Syria	Yes	No	0.54 (residential <100 kWh/month) 0.76 (100–200 kWh)	8.69 (residential/commercial/ industry >2000 kWh)	N.A.
Saudi Arabia	Yes	No	1.33 (residential <2000 kWh/month	6.91 (residential >10000 kWh/month)	No
Yemen	Yes	No	2 (urban residential and mosqs <200 kWh/month	9.05 (government)	No
Oman	Yes	Seasonal for industry only	2.6 (0–3 MWh; 0–7 for agriculture, fisheries)	7.8 (residential and government >10 MWh)	Considered
UAE (Abu Dhabi)	Yes	No	0.8 (farms)	4 (all customers except UAE nationals)[b]	Considered
Qatar	Yes	No	1.9 (Industries)	4.1 (government)[c]	No
Bahrain	Yes but only for residential	No	0.8 (residential <3000 kWh)	4.24 (residential >5000 and all non residential)	No
Kuwait	No	No	0.3 (subsidized industries)	4.4 (non-subsidized industries)	N.A.
Iran	Yes	No	0.01	4.2	Considered
Iraq	Yes	No	0.84 (all <1000 kWh)	4.19 (all >4000 kWh)	N.A.

Note: Tariffs are in $ cents/kWh for 2010, except Israel (2007) and Iran (2005)
Source: Local utilities and the Arab Union of Producers, Transporters and Distributors of Electricity
SMI small and medium industries
[a]90% of residential sales are for less than 350 kWh/month
[b]UAE nationals pay $1.36 cents/kWh
[c]The second highest tariff is 3.8 paid by commercial sector and hotels for more than 15,000 and 500,000 kWh/month respectively

Table 7.5 Electricity tariffs and direct subsidies in 2003 in Iran (Rials/kWh)

Consumer category	Supply cost	Electricity selling price	Subsidy as % of set price
Residential	407	97	76
Public	318	152	52
Agriculture	323	14	96
Industrial	289	163	44
Commercial	396	412	
Average	315	132	60

Source: Tavanir (2003–2004); $1 = 9928 Iranian Rials (September 12 2009)

7.3 Summary

The credibility of the reform—or its capacity to provide stability of commitments, and the ability of its governance structure to solve regulatory problems between the government, the operators and the other stakeholders—is critically supported by the design of "effective" regulatory institutions, meeting a set of criteria. In MENA, clarity of roles, participation of relevant stakeholders and accountability are usually guaranteed. The transparency of the decision-making process is often fulfilled by economic regulation and open bidding procedures, except under specific circumstances. The autonomy from political intervention and the predictability of decisions are less certain. More concretely, the design of effective regulatory institutions hinges on the availability and the use of an adequate supply of economic resources, particularly of human resources. The development of robust regulatory frameworks and strong institutions was hampered by under-funding and a reluctance of authorities to transfer independent decision-making to regulatory authorities.

The credibility of these reforms could not be analysed without looking at the subsidies issue and pricing of energy. The growth in the volume of subsidies is usually not offset with the hypothetical decrease in production costs due to efficiency and competition. A subsidy applied through fuel pricing can potentially be counterproductive to stimulating efficient use of energy. Increasing fuel prices would encourage generators to invest in energy efficient technologies. The restriction of fuel supplies obliged most countries to rethink their position vis-à-vis subsidies. Regionally, the value of natural gas is becoming increasingly linked to, at least, its cost of production or to its export netback value. But there are still several constraints affecting the review of subsidies which transcend the political difficulty such as macro-economic policies, lack of alternatives, etc. In order to achieve a minimum level of credibility, authorities are encouraged at least to, first, have a clear idea of their energy subsidies and, second, to make their amount and the calculation methodology transparent (Egypt, Abu Dhabi, Oman).

Chapter 8
Evaluation Step 5: Remediability, or the Critical Path to Survival

Abstract The test of remediability of Williamson enabled us to conclude that, in the 12 countries which reformed their electricity sector, the Single Byer Model (SBM) appears to be a relevant model so far. But the growth of consumption centres, of distribution companies, and of large industrial centres, with interest to sign bilateral contracts with generators, is a reality. An "SBM plus" (an SBM model where bilateral contracts could be signed) could be one *feasible alternative*. Market power could be a major impediment for remediability. A well-connected, but still independent, national champion is not necessarily a bad outcome, as long as regulation is strong.

Keywords Contestable market · Bilateral contracts · Single buyer model

It is important to have defined the analytical framework built on the first three criteria (attractiveness, feasibility and credibility). Experiences show that stability of engagements is not sufficient to guarantee the credibility of a reform. They also indicate that the ideal model, which could be duplicated to several countries, does not exist. In this context, it remains important to maintain some level of *flexibility* in order to allow improvement. This enables us to use the test of remediability of Williamson to evaluate the capability of the reform to keep open alternatives of enhancement. Williamson (1996: 195) states that *«an outcome for which no feasible superior alternative can be described and implemented with net gains is presumed to be efficient»*. The feasibility and the implementation with net gains are central, as technological, political, institutional and administrative transformation will come at a cost, when feasible. However, it is critical to evaluate the feasibility as this argument of "relative efficiency" could be used to maintain a long-lasting status quo or to avoid profound reforms (Haggarty and Shirley 2002; Williamson 2002).

Perez (2002) analysed the remediability criteria for three European experiences: the British, the German and the Spanish experiences. He highlighted the linkages with the coordination capabilities of the authorities and with the intensity of the negotiations' constraints. He concluded that the British reform (of 1990), with an important institutional and industrial feasibility, does not fulfil though the

L. Benali, *Electricity-sector Reforms in the MENA Region*, Perspectives on Development in the Middle East and North Africa (MENA) Region,
https://doi.org/10.1007/978-3-319-96268-9_8

remediability criteria, because of the choice of duopolistic industrial structures.[1] In Spain, with an important institutional feasibility and a somewhat limited industrial feasibility, the remediability criteria cannot be used to justify the creation of national champions (namely Endesa and Iberdrola). The government does not appear to desire profound reforms. The limited institutional and industrial feasibility of the German experience enabled surprisingly a performing auto-regulation. Paradoxically, the complexity of the institutional political set-up in Germany (federal structures, judiciary supervision) and the industrial modes of the electricity sector (initial vertical and horizontal unbundling, subsidies' reduction to coal) created one of the most dynamic reforms in Europe.

Out of the 18 countries in the MENA region, six have not introduced or have failed to introduce competition in the generation.[2] The remediability of the reforms in the remaining 12 countries is analysed using the following criteria grid:

- The creation of national champions or the strengthening of existing ones
- The effective move towards more deregulation, and whether the attempt was successful
- Institutional and Industrial feasibility
- The concentration of generation and the size of each market in mind (Table 8.1)

These 12 countries have introduced various forms of the single-buyer model, to relieve capacity shortages while conserving public resources. They authorized private investors to build or buy power plants to generate electricity and sell it to a single buyer—usually the national utility but not always—. IPPs sell their output through long-term power purchasing agreements, sometimes with take-or-pay clauses or fixed capacity charges to protect investors from market risks. Most proceeded to the unbundling of the national utility into separate departments or companies for generation, transmission and distribution. Generation and distribution companies would be the candidates for privatization, once they are corporatized and restructured. Most countries awarded exclusive rights to the transmission and dispatch company (or what remains from the national utility) to purchase electricity from generators and sell it to the distribution companies. Unbundling allows separating generation from transmission and distribution and facilitates competition by enabling equal treatment between private and government-owned generating assets. Indeed, the most straightforward way to ensure this equal treatment is to have them sell their output to the single buyer and rule out direct bilateral contracts with distributors, or even with large customers. Lovei (2000) details the technical, economic and institutional reasons behind the popularity of the single-buyer model. He explains that, as electricity supply has to match demand on a real-time basis (second by second), a single entity responsible of real-time dispatch and of procurement and selling of electricity makes it easier to balance differences between planned and actual generation and demand. A single buyer model does not need

[1]The duopolistic industrial structure created situations of intense market powers (Newberry 1998).

[2]See Chap. 1, Sect. 2.b.

Table 8.1 Market concentration and remediability in reformed MENA electricity markets

	National champions	Move towards Deregulation/ successful if so	Institutional feasibility	Industrial feasibility	Share of government-owned entities in IPPs	HHI	Installed capacity 2006 (GW)
Morocco	Yes	Yes/No	Medium	High	0–48%	3090	5
Algeria	Yes	Yes/No	Low	Medium	49%	9414	9.3
Tunisia	Yes	No	High	Medium	0%	6326	2.8
Egypt	No	Yes/No	Medium	Medium	0%	7628	17.7
Jordan	No	No	High	High	0%	6226	1.9
Lebanon	Yes	No	Low	Medium	40%	8166	2.3
Saudi Arabia	Yes	Yes/No	High	Low	40% minimum	7324	32
Oman	No	No	High	High	0%	3656	3
UAE (Abu Dhabi)	Yes	No	High	Medium	60%	1565	18 (UAE)
Qatar	Yes	No	High	High	45–60%	4570	3.4
Bahrain	Yes	No	High	High	0%[a]	3513	1.4

Source: local sources, author's calculations
[a]The Gulf Investment Corporation (GIC), equally owned by the six GCC countries, has shares in two out of Bahrain's three IPPs

Third party access to transmission, which can be costly and complex to implement, and reduces market and regulatory risk and therefore financing costs. Price regulation is simplified (unified wholesale electricity price). But more importantly, the institutional feasibility of the Single buyer model is high as ministries keep a key role in major investment decisions and the national utility in day-to-day oversight. In a nutshell, the Single Buyer Model offers technical, regulatory and institutional simplifications, while preserving some of the State's control of wholesale electricity trading.

As a matter of fact, the SBM means control: technical control, contractual control, economic control and institutional control, and therefore less "anxiety". Why would a national planner be ready to lose the benefit of this control? The status quo reached in various countries leads us to explore whether this situation would reflect a sort of Nash Equilibrium, assuming intelligent rational behaviour of all players involved in the decision to reform or not to reform (government, utilities, consumers or their representatives including industries, emerging players, international donors).

Examples of behaviours include: International donors would assume that the SBM could only be temporary and could only lead to further deregulation. In contrast, governments would still want to see the performance of the sector increased but not at the expense of the supply of power at peaks and extreme peaks. Governments would not be ready to make concessions on peak capacity adequacy even if it means intervening directly in the sector by turning IPP projects into EPC projects, or by imposing reserve obligations. A less interventionist approach, but none the less centralized, would consist in the procurement by auctioning for peak capacity. Even in deregulated markets, markets' capacity-regulating mechanisms are necessary. Even so, they can cause significant price, reliability and security pressures that could affect the market rules and that need to be managed. Producers would want higher prices, and it is increasingly noticeable in recent bids that prices offered in tenders are following that trend. This could be partially explained by the escalation of costs. However, the changing structure of the industry with more consolidations and coalitions is also supporting prices.

A Cournot-Nash equilibrium analysis would have been only partly helpful to explain the behaviours of players in a restructuring regime. The fact that these players are institutional players rather than firms confront us to the predictive limits of economic analysis, because the outcomes can not be completely predicted given the various and changing drivers of each category of players. The theory of non-cooperative game could enable to analyze for each market the ideal number of firms and the associated fuel mix in generation to optimize the output price function. Applying a computable general equilibrium (CGE) model to specific countries in the MENA region would have been interesting, and this is nascent. In that sense, the assumption of perfect rationality (optimal pricing) could be used without too much distortion on the end results. However, it does not answer the core question here, i.e. whether the SBM in the region constitutes a Nash equilibrium that maximizes the expected utility pay-off of each player (in this case, the institutional players), against the strategies of all the other players. The assumption of intelligent rational behaviour means that the behaviour of all players in such a game can be predicted. Short of that, there should be at least one player whose expected utility pay-off or welfare could be improved, by re-educating him to pursue his interest more effectively. As an example, some customers would want stability of supply and guaranteed prices, preferably via long-term contracts or via the construction of their own captive plants (large electricity intensive industries). These categories of consumers would push for a "return" to vertical integration. They would prefer to either sign long-term guaranteed supply contracts disconnected from the spot price (if any) or from the wholesale tariff, or to build their own captive plant, with the possibility to sell the surplus to the grid if they are allowed to. Actually, what Finon (2004) considers as a "secondary route" only (long-term contracts to avoid price volatility risk) is very relevant in the case of several Middle Eastern countries who embarked in ambitious industrial plans.

Lovei attempts to list the downsides of the SBM and states that it is better to skip the transitional arrangement that could be the SBM and adopt right away a model with multiple buyers. For Lovei, the downsides are mostly linked to countries with "weak or corrupt government and low payment discipline". The first drawback raised relates to over-building of capacity in case of available state guarantees and favourable market conditions. Second, PPAs create a contingent liability for the government, which usually does not appear on governments' budgets. Third, the SBM would respond poorly in case of lower than expected electricity demand. Instead of having a price decrease which would stimulate demand, wholesale electricity prices would rise because of take-or-pay clauses or fixed capacity charges. Fourth, Lovei argues that the SBM hampers the development of cross-border electricity trade, and fifth that it weakens incentives for distributors to collect payments—if paying and non-paying distributors are treated alike—. Sixth, governments intervention in the dispatch of generators and the allocation (or worse, the diversion) of cash proceeds is a risk. Finally, governments could maintain the status quo and indefinitely delay the next step in the reform process toward fully liberalized markets.

The main caveat in Lovei's analysis is that he assumes a weak or absent regulation and seems to offer a critique of some empirical applications of the SBM instead of the theoretical model itself. Moreover, he also presumes from the outset that the SBM could only be a transitional arrangement and that the ultimate goal for all countries would be to fully liberalize their electricity sectors. Finon 2004 *highlights that the limits of the SBM are actually more related to the environment in which it is deployed than with its intrinsic downsides.*

On the other hand, Lovei advocates direct bilateral contracts between generators and consumers, as a more suitable model for developing countries than a mandatory competitive pool.[3] They diminish the governments' ability to intervene in

[3]"The Ukraine experience with a mandatory pool—with the government intervening arbitrarily in the allocation of cash proceeds and generators unable to stop supplying electricity to delinquent distribution companies—shows that the disadvantages can have severe consequences". Moreover, in many developing countries the electricity system may be too small for a pool to function competitively.

investment decisions and in the payment chain, but they also pose challenges that must be tackled upfront. These include the need for the creation of a *balancing mechanism* to maintain equilibrium between supply and demand on a real-time basis, the need to develop a *transmission access and pricing regime*, reflecting capacity constraints and loss factors, to support the transactions, the *suboptimal dispatch* schedules and the absence of a unified wholesale price therefore the need to spell out *procurement rules* for distributors before they can pass through costs to consumers.

According to Lovei, "the short-term costs of developing solutions to these challenges are worth the investment, given the compelling long-term benefits of the bilateral contracts model: reduced likelihood of government interference, incentives to improve payment collection, market-driven decisions on new capacity, and better opportunities for cross-border electricity trade." Weighing short-term real costs against long-term hypothetical benefits is a core issue.

Hunt (2002) rules out the retail competition model as needlessly expensive and complex, leading to mistakes and aggravation.[4] She states that experience shows few small customers exercise their choice to change contracts, and that the same results could be achieved with a choice of tariffs (instead of a choice of suppliers/providers). One last complication mentioned is that distribution companies would have to buy in auctions should they have to serve small customers.[5] A monopolistic structure, despite its drawbacks as mentioned earlier, is "often better than a bad restructuring". Private monopolies do not necessarily perform better than government-owned monopolies (case of France). They are sometimes used to shift politically unpleasant tasks out of government but they create an important interest group which might impede future liberalizations if any. Moreover, they would not necessarily support the country's major macro-economic and energy policies.

If there is willingness to move away from a monopolistic structure and to introduce some competition, wholesale competition appears indeed to be the safest bet, assuming the market is large enough. Moreover, take-or-pay clauses in long-term PPAs are less of a burden for single buyers in environments of high demand growth (the likelihood of large overcapacity is slim). However, most countries in the region stopped at the Single Buyer model, as they introduced wholesale competition. One country (Morocco) tried to deregulate further but has been struggling with the implementation. Is this a signal that electric reforms in the region do not fulfil the remediability criteria as defined by Williamson? Or simply that no feasible alternatives to the SBM exist?

Feasible alternatives exist: the Multiple Buyers Model in its various forms (large customers, distribution companies, spot markets, utilities…etc). With the Single Buyer Model, it became common that companies were willing to bid for projects if there were long-term government guarantees, and the risk remains on the government. There are a few exceptions in the region of entities which have good enough

[4]See Appendix B for a description of the four organizational models in the electricity sector.

[5]Examples include Columbia, New Jersey and Brazil.

credit to sign long-term contracts, like the Saudi Electricity Company (SEC), an internationally rated company (AA− on July 2009, A+ in April 2016—Fitch) and does not need to carry government guarantees. In the case of the Multiple Buyers Model, the risk is usually transferred to the generators. Power Purchasing Agreements would more probably continue to dominate, but they can be reduced in time length. However, in comparison with the SBM, the MBM requires additional arrangements as far as transmission access and trading are concerned. It obviously needs a multiplicity of buyers but also a well-managed transmission grid, a competition law and a solid infrastructure. In general, wholesale competition should be well-implemented to allow further progress of future reforms if need be, but this is all the more relevant when buyers and transactions become multiple.

Does this alternative need a spot price to settle the contracts? Not necessarily. Given that MENA countries are facing rapid and uncertain growth in demand, it is important that the model implemented is designed around the need to install enough capacity (generation and T&D) on time. This is generally the case in non-mature markets. *Long-term coordination and bilateral (long-term in a first instance) contracts* are the necessary conditions whichever model is chosen. They would allow having a centralized planning of investments, based on a "bottom-up" approach of demand. They would provide a "guarantee" of investors' response given that investors would see limited risks (allowing them to raise the necessary finance more easily).

A few countries have introduced variations in the SBM, to allow some flexibility and capture some of the advantages of the MBM but without fully engaging in it. This is a signal that they start to see the advantages, and the feasibility, to move to a variant of the SBM. Moreover, countries have to prepare themselves for the long-term possibility of a deceleration of demand growth.

Exemption orders (like in Oman where they are issued by the regulator) could be used in order to exclude from the SB purchase duty, and to allow upstream operators or large industrials for example to generate their own electricity. In some cases, captive generators could inject their surplus electricity in the grid. However, that would require an adequate and robust transmission system and it would bear technical risks (even risk of blackout as happened in Bahrain during the summer of 2004, see Sect. 2 of Chap. 1). Ad-hoc memorandum of agreements, within the framework of the license attributed by the regulator, could also be used.[6] However, in cases of market power, it is highly likely that these mechanisms would benefit the dominant player and strengthen its position.

As a matter of fact then, there might be mechanisms to maintain some flexibility away from the SBM, but it is still not the case that a yet-to-be-defined "eligible" customer could procure electricity from any, or at least from more than one, generator.

---

[6]In July 2009, Knowledge Economic City (KEC) signed a memorandum of agreement with Saudi Electricity Company (SEC) for the supply and distribution of power to the city.

The "SBM Plus" is defined as an enhanced Single Buyer Model which allows to-be-defined eligible customers to sign bilateral contracts with suppliers.[7] For the SBM plus to function optimally, one necessary condition is to have a multiplicity (or at least more than one) of credit-worthy entities which could sign these contracts without the need to resort to sovereign guarantees. This multiplicity will give eligible customers the possibility to choose between suppliers, and might enable the contracts to be shorter.

One major potential impediment for the move to a potentially enhanced model, or to remediability, is market power. In most countries in the region, it is the historic player which might continue to hold market power as explained earlier. Some, mainly in the Gulf, have tried to break dominant positions but then, the temptation to create (or re-create) national champions (particularly those which could invest overseas) is great. Built on the experiences of Morocco, Saudi Arabia, Oman and Abu Dhabi particularly, it appears that the preliminary solution to solve this dilemma could be: If a national champion is needed or desired, it is advised to insulate it from local politics, e.g. by staffing with result-oriented technocrats or expatriates. Moreover, there could be situations where breaking up monopolies without giving to the new entities the full means to develop could be counterproductive.

The concept of the dual market model of dual firms was defined earlier. As a reminder, this would be a high-profile politically well-connected (to mobilize resources that are essential to their commercial viability) and sufficiently well managed firm. A well-connected, but sufficiently independent, national champion provides an attractive and feasible solution to the immediate concerns of the electric sector (performance, delivery and financial viability). In this case, as mentioned earlier, strong regulation is important. However, attractiveness and feasibility would be fulfilled at the expense of remediability.

On the other hand, the need to maintain the "sacred" consumer protection should play a balancing role and guarantee the remediability of the reforms. Consumer protection is probably as important as power and might protection given the implicit "social contract" between the authorities and the population. Moreover, the potential success of the industrialization policy was built upon the notion of protection of industrial customers—against abusive behaviour concerning fuel or power supply. As in other markets, electricity consumers require protection against adverse industry behaviour. Such behaviour may be *economically rational* but *objectionable* from the perspective of justice and fairness. Since consumer protection may have been implicit under monopoly provision, restructuring requires establishing explicit rules for minimum customer service standards which electric utilities must respect when providing electric service to the public. This includes articulating rules for payment of bills, late payment and non-payment. Rules are also needed to define who is eligible for special treatment, when they are eligible and for how long. For example, special dispensations may be justified for such factors as medical conditions, age

[7]Similar models exist in some gas markets, but gas does not have the same technological constraint as electricity (transmission).

(elderly or infants in the home) and times when temperature conditions are extreme. More lenient contractual or financial terms may also be justified. For example, it may be reasonable to waive security deposit requirements and certain fees if these have the effect of excluding certain households from accessing electric service. But rules would be powerless without mechanisms and entities to reinforce them. Therefore, beyond the establishment of rules, bearish authorities would often introduce "remediability clauses" to be able to reverse a reform move if consumer protection is threatened (including if power projects risk being delayed).

At this stage, there are three preliminary conclusions:

- The SBM appears to be the most relevant model for most countries in the region, but the status quo should not be maintained because of the apparent absence of alternatives. Given the growth in consumption centers and the emerging challenges facing the region, there are several places where an "SBM plus" make sense, i.e. where bilateral contracts could be signed, assuming all the conditions to implement these contracts in the most efficient way are met (existence of more than one credit-worthiness entity to keep government away from contract and to give customers choice, access to transmission, robust network system and demand management). The "SBM plus" is one feasible alternative.
- A major impediment for remediability would be market power. In that sense, most electric reforms in the region do not fulfill Williamson's remediability criteria. Even in those countries which tried to break up monopolies and shift some power away from incumbent, the temptation to create (or re-create) national champions is great. Breaking up monopolies without giving the new entities the full means to develop is obviously counterproductive. At the end of the day, if a national champion is needed or desired, it is preferred to keep politics away from it. In this regard, a well-connected, but sufficiently independent, national champion provides an attractive and feasible solution to the immediate concerns of the electric sector (performance, delivery and financial viability). Strong regulation will be important, recognizing that attractiveness and feasibility would be fulfilled at the expense of remediability.
- However, the need to maintain the "sacred" consumer protection should play a balancing role and guarantee "some" remediability of the reforms. Consumer protection is probably as important as power protection given the implicit "social contract" between the authorities and the population. Moreover, industrialization policies are built upon the notion of protection of industrial customers –against abusive behaviour concerning fuel or power supply.

8.1 Interim Conclusion

Today, economics cannot be narrowly defined as a social science concerned with the production and allocation of material goods anymore. Rather, it relates to incentives in several forms of social institutions. The evaluation grid developed here could be applied to other types of reforms: social, macroeconomic, political, etc. It could particularly be used when benefits and losses are difficult to quantify. Evaluating electricity reforms in developing countries suffer from lack of transparent data. The absence of retail price signals because of subsidies can cloud the picture if wholesale costs are not accessible, and even there demand response is often overlooked and marginalized in analysis. In the absence of a direct link between generation costs and retail prices, the quantitative analysis attempted to analyse the impact of electricity reforms in the MENA region on costs and investments. The impact on costs is unclear, but at least, one could conclude that electricity reform did not lead to a substantial decrease in costs (comparable in the case of Abu Dhabi for instance to those of 1999) even if expectations of economic performance played a critical role in launching these reforms in the first place. The RPI-X regulation is acting as an efficiency incentive for most generators and government-owned assets will be gradually in competition with the others. The impact on investments is more clear-cut: electricity reforms contributed to attract some investments in the region but failed to cover all investment needs. For instance, the "fuel offer", embedded in long term "$1-gas" supply contracts, made the region attractive for power developers but it was not sufficient, and more importantly, it was not sustainable. Future quantitative evaluations will be even more difficult as fuel prices are being be reviewed and recent inflation of capital costs will come through the system. The role of institutions which can perform this evaluation exercise, ideally independent regulators or other entities, becomes even more critical. That would assume however access to relevant data and information and availability of resources to conduct the evaluation.

In MENA, countries reformed their electricity sectors differently. Even if all went for the Single Buyer Model, they chose different paths to reach it. The paths are diverse in the number of steps taken, the length of the process and the sequence of the steps. Long reforms were not always the most successful but quick reforms bear the risk of being defective. Actually, long reforms can be interpreted as the sign of rigid industrial structures and complex institutional set-ups, and might not solve the issue of market power. The region still displays an important concentration of players in the generation of electricity. Highly concentrated markets, can have some advantages (similar to the advantages offered by monopolies) in some cases if they are properly regulated, but their sustainability can be questioned. To be sure, the attractiveness, the feasibility, the credibility and the remediability of these reforms can be used as qualitative evaluation metrics.

The concept of **attractiveness** (economic), with the desirability (political), is central to the analysis of reforms in the neo-institutional economy. The attractiveness and desirability of a reform are usually driven by three main factors: an economic crisis, a sectorial crisis or a major political change. However, this theory, relevant in the case of reform leaders like the UK, disregards the impact of "exogenous factors" and does not appear relevant in the case of most countries in the Middle East and North Africa. All reformers in the region were neither driven by a sectorial crisis nor by a major political change. In some countries, economic crisis might have probably been the most relevant factors in the pre-2000 period, but surely not afterwards. The history of electricity reforms in the region shows that the attractiveness and desirability was mainly driven by other factors which do not fall necessarily in the three above-mentioned categories. The focus has tended to be on the introduction of competition and exogenous factors (international trends...) were important in that process. The impact on macro-economic policies (sought industrialization, economic diversification) is a key component of the attractiveness of MENA reforms.

The institutional feasibility relates respectively to the degree of centralization of powers and decision-making. The industrial feasibility "measures" how easily transformable the industrial structure could be. The more decentralized the institutional environment is, the less feasible the reform will be without negotiation and spending of political capital. The more rigid the industrial structure is, the longer the reform will be. Time is a critical factor as it increases the probability for the reform to be distorted, reversed or frozen. However, transaction costs would be higher in countries with a rigid industrial structure, and it would be more difficult to reverse or to freeze a reform in an advanced stage.

In a region where world-class oil and gas reserves are concentrated, a feasibility analysis would not have been complete if it did not look at how the resource endowment of a country affects the feasibility of a reform, and its process. The case of Saudi Arabia is particularly unique in the way its resource endowment has been influencing the reform of its electricity sector.

The **credibility** of the reform—or its capacity to provide stability of commitments, and the ability of its governance structure to solve regulatory problems between the government, the operators and the other stakeholders- is critically supported by the design of "effective" regulatory institutions, meeting a set of criteria. In MENA, clarity of roles, participation of relevant stakeholders and accountability are usually guaranteed. The transparency of the decision-making process is often fulfilled by economic regulation and open bidding procedures, except under specific circumstances. The autonomy from political intervention and the predictability of decisions are less certain. More concretely, the design of effective regulatory institutions hinges on the availability and the use of an adequate supply of economic resources, particularly of human resources. Most gulf countries have been able to attract high calibre international economists. Other countries in the region, particularly in North Africa, struggle to attract the few human resources they had. In some cases, the development of robust regulatory frameworks and strong institutions was hampered by under-funding and a reluctance of authorities to transfer independent decision-making to regulatory authorities (even when required to do so by Law). Transitional arrangements are needed (initial price controls in contracts or licenses or clear tariff setting principles in legislation), but enhancing regulation should be a continuous and evolving process. The reality is that this is easier in some countries in the region than in others.

Some (Morocco, Lebanon, Bahrain) decided to engage in a reform process without the creation of a regulation authority. This option bears cost and risk advantages for the State, but in general, an effective regulation in the absence of a regulatory authority assumes that the controlling role of local collectivities and professional associations is reinforced. The regulation would be non-discretionary, usually ex-post (while a regulatory authority would regulate ex-ante), less information-intensive, discontinuous and dependant vis-à-vis the political framework. In a context of vertical integration, network managers tend to deploy opportunistic behaviours to indirectly limit competition.

Regional experiences show that the coexistence of a strong State and a dependent (sometimes weak or absent) regulator did not prevent the move to the single buyer model (SBM) and the introduction of competition. In fact, it sometimes expedited the process. In the future, governments might be less willing to give sovereign guarantees if competition for projects increases, and if some competitors can find other ways to mitigate the risks without transferring them entirely to the State. Successful IPPs in the region were built on the credibility of the entities, on the predictability of decisions and on the clarity of the processes.

The credibility of these reforms could not be analysed without looking at the subsidies issue and pricing of energy (electricity and fuel, as in several MENA countries, the electricity sector is subsidized by applying a "double subsidy"). When competition is introduced and electricity demand is growing, the growth in the volume of subsidies is usually not offset with the hypothetical decrease in production costs due to efficiency and competition. A subsidy applied through fuel pricing can potentially be counter-productive to stimulating efficient use of energy. Increasing fuel prices would encourage generators to invest in energy efficient technologies. It would also increase the cost of electricity, hence increasing the level of direct subsidies to the electricity sector, assuming the same retail price. the simulation model enabled to compare wholesale prices with theoretical generation costs under different assumptions for projects locations and fuel costs. The restriction of fuel supplies obliged most countries to rethink their position vis-à-vis subsidies. Regionally, the value of natural gas is becoming increasingly linked to, at least, its cost of production or to its export netback value. A few countries have started to negotiate more sophisticated terms (gas prices indexed to end-user pricing or including a sharing of end-product netbacks) in recent contracts with some industries. But there are still constraints affecting the full lifting of subsidies which transcend the political difficulty such as macro-economic policies, lack of alternatives, etc. In order to achieve a minimum level of credibility, authorities are encouraged at least to, first, have a clear idea of their energy subsidies and, second, to make their amount and the calculation methodology transparent. A few countries have started doing that (Egypt, Abu Dhabi, Oman and more recently Saudi Arabia). Others will have to follow suit should they want to guarantee that minimum level of credibility for their electricity reforms.

Finally, as the ideal model, which could be duplicated to several countries, does not exist, it appears important to maintain some level of flexibility in order to allow improvement. The test of **remediability** of Williamson[8] enabled us to conclude that, in the 12 countries which reformed their electricity sector, the SBM appears to be a relevant model so far. Actually the analysis of the advantages and shortcomings of the SBM as it applies to the region shows that its limitations are more related to the

[8]To evaluate the capability of the reform to keep open *feasible alternatives* of enhancement.

institutional and macroeconomic environment in which it is deployed than with its intrinsic weaknesses. But this is not an excuse to maintain a long-lasting status quo. The growth of consumption centres, of distribution companies, and of large industrial centres, with a potential interest—cost or volume—to sign bilateral contracts with generators, is a reality. Therefore, there are several places in the region where an "SBM plus" could be one *feasible alternative*. The SBM plus is defined as an SBM model where bilateral contracts could be signed. The conditions to implement these contracts efficiently include the existence of more than one credit-worthiness entity to keep government away from contract and to give customers choice, access to transmission, robust network system and demand management.

Market power could be a major impediment for remediability. In that sense, most electricity reforms in the region do not fulfil Williamson's remediability criteria. Even in those countries which tried to break up monopolies and shift some power away from incumbent, the temptation to create (or re-create) national champions is great. Moreover, breaking up monopolies without giving the new entities the full means to develop is obviously counterproductive. A well-connected, but still independent, national champion is not necessarily a bad outcome, as long as regulation is strong. It appears that market concentration is not in itself an impediment for a certain level of competition in the sector. Percebois (2008) affirms that the important point for the market is to be contestable (Chicago school approach and the antitrust paradox), but the regulatory or overseeing authorities will have to ensure that operators do not adopt strategic behaviours aiming at raising prices.

It is clear however that attractiveness and feasibility of these pragmatic solutions would be pursued at the expense of remediability. The need to maintain consumer protection (citizens, industrials) should play a balancing role and guarantee "some" remediability of the reforms.

8.2 Summary

As the ideal model does not exist, it is important to maintain some level of flexibility in order to allow improvement. The test of remediability of Williamson enabled us to conclude that, in the 12 countries which reformed their electricity sector, the SBM appears to be a relevant model so far. Actually the analysis of the advantages and shortcomings of the SBM as it applies to the region shows that its limitations are more related to the institutional and macroeconomic environment in which it is deployed than with its intrinsic weaknesses. But this is not an excuse to maintain a long-lasting status quo. The growth of consumption centres, of distribution companies, and of large industrial centres, with interest to sign bilateral contracts with generators, is a reality. There are several places in the region where an "SBM plus" (an SBM model where bilateral contracts could be signed) could be one *feasible alternative*. The conditions to implement these contracts efficiently include the existence of more than one credit-worthiness entity and to give customers choice, access to transmission, robust network system and demand management.

Market power could be a major impediment for remediability. Most electric reforms in the region do not fulfil Williamson's remediability criteria. Even in those countries which tried to break up monopolies and shift some power away from incumbent, the temptation to create (or re-create) national champions is great. A well-connected, but still independent, national champion is not necessarily a bad outcome, as long as regulation is strong. The important point for the market is to be contestable, but the regulatory or overseeing authorities will have to ensure that operators do not adopt strategic behaviours aiming at raising prices. The need to maintain consumer protection (citizens, industrials) should play a balancing role and guarantee "some" remediability of the reforms.

Chapter 9
Prospects and Future of Reforms: Fuel Mix Options

Energy companies do not determine the energy mix, governments do, through granting permits, imposing or lifting taxes and providing subsidies. The energy mix is a government decision
Jeroen Van der Veer, former CEO Royal Dutch Shell, the 10th International Oil Summit, April 2009

Abstract Electricity reforms are not in contradiction with a fuel diversification policy, but introducing new fuels and new technologies (renewables for instance) would require the State to intervene directly or indirectly. Regulation will play a critical role and changes in fuel offerings will require a delicate balancing act. They will test all already established market models in the region and will challenge decisions to deregulate further. Out of the 18 countries in MENA, 12 are hydrocarbons exporters (8 of which are OPEC countries). Hydrocarbons dominate the fuel mix, and natural gas has been considered as the fuel of choice for power generation. But natural gas will need to be complemented by other fuel sources. In some countries, oil-fired generation continues to prevail, while other alternatives are considered: renewables, nuclear and coal. Evolving fuel prices could be viewed as challenging for any introduction of an additional degree of deregulation. The transition is meant to accompany the country as it transforms from a fuel exporter to a fuel monetizer and then to a mature economy.

Keywords Fuel mix · Renewables · Nuclear · Natural gas · Oil

This book attempted to build a tool to evaluate reforms, with an application to the region. One main emerging conclusion seems to be that the Single Buyer Model, or the introduction of competition at the wholesale level, will remain relevant for the foreseeable future. Of course, there are challenges inherent to the system. The independence of regulators in relation with other institutions needs to be reinforced at various levels. A new market structure would lead to elevated management requirements and an increasing need for coordination. The various contractual

L. Benali, *Electricity-sector Reforms in the MENA Region*, Perspectives on Development in the Middle East and North Africa (MENA) Region,
https://doi.org/10.1007/978-3-319-96268-9_9

agreements (PPAs, FSAs) will need to be self-consistent and aligned with the prevailing legislation. The SBM could create (or recreate or strengthen) powerful companies which would be free from political control. It may be about time for some of these countries to consider an alternative, although in most cases, market power has already significantly developed to prevent future reforms. In this regard, the analysis would not be complete without confronting the future of the reforms in the region to a set of policy decisions and orientations that could be, arbitrarily or otherwise, chosen by governments.

Indeed, the region is facing new challenges, issues and opportunities which could have an indirect impact on reforms:

- There are growing interconnections and interdependencies between countries, mainly for electricity, less for gas. It is all the more relevant that the different parts of the energy chain are becoming more integrated. For example, it is becoming increasingly difficult to plan the power sector in complete isolation of the upstream sector, when the two sectors are increasingly interdependent.
- There are several new types of players involved in the energy sector. Some of them were historically active in the E&P business. Others, like Asian companies which were virtually prevented from participating in the power sector in the region, eventually found their way in.[1] Finally, local state-owned entities (different than national utilities), banks, funds, private companies are increasingly investing in power assets inside and outside of the region.[2]
- The diversification of the fuel mix is becoming high on governments' agendas for different reasons (environment, strained fuel supplies, geopolitics). This poses the important question of fuel pricing and potentially of consumers' electricity tariff evolution.
- Countries are increasingly concerned about the need to achieve a "sustainable" development growth, while maximizing/optimizing the use of their natural resources.

These policies behave like exogenous factors, vis à vis the electricity sector. They could hamper, encourage or not impact at all the ongoing reforms in the region. They are exogenous as they could be followed regardless of the electricity reform movements in the specific country. For the sake of clarity, this analysis opted for the last two sets of policy orientations: electricity fuel mix options and resources monetization. Also, it we focused on hydrocarbon-exporting countries in the region, assuming

[1]Asian countries are becoming important players in providing services, sometimes technology and even investment to build generation plants or other electricity projects, as part of their effort to improve ties with Middle Eastern countries. With special financing and other incentives on offer from Asia, and in some cases fuel to supply the power sector, Middle Eastern countries could have limited incentives to establish open competitive power markets, with the rise of Asian players.

[2]These new local actors investing in power assets include examples like TAQA and Liwa Energy (Abu Dhabi), Oman Oil Company and Gulf Investment Corporation.

that the behaviour of MENA hydrocarbons' importing countries would be similar to other energy importing countries in other regions in the world.

> The energy problem can't be left to pure capitalism, governments and companies need to cooperate
>
> Jeroen Van der Veer, former CEO, Royal Dutch Shell

In all countries globally, the choice of an electricity fuel mix has been a critical aspect shaping the power sector. Indeed, the behaviour of the government towards reforms and the introduction of competition are completely different when the fuel mix is dominated by nuclear (France), coal (South Africa), oil or gas (most MENA countries). Another more subtle layer of differentiation lies in the question whether fuel resources are endogenous or imported. Access to fuel (volumes and costs) is then a major component of governments' "marketing offer" when they are willing to introduce competition and attract the right players. This is of course very relevant for energy importers, but also for energy exporters since international markets are becoming more liquid and fuel export netback prices, or opportunity costs, are increasingly being viewed as the right benchmark for energy pricing.

For strategic and geopolitical reasons, nuclear has been often seen as a prerogative of the State (see section on nuclear). Hydrocarbons and coal are more "commoditized" but for hydrocarbon exporters, oil- and increasingly gas-utilization in the electricity sector has to be aligned with the country's overall export policy. In the region, most countries are revisiting their attitudes toward an electricity fuel mix dominated by hydrocarbons, arguing the need to free up more hydrocarbons for exports. But the real issues related to the fuel mix usually lie elsewhere—in the need to limit burning valuable distillates and the rising competition for an increasingly expensive natural gas. For current and future energy importers in the region, the need to ensure some level of security of supply is also relevant.

Another area where linkages between different fuels play a role is in OPEC countries, where there is a need to maintain a proper balance between oil exports and domestic consumption. The power sector could be then used as an outlet for discounted crudes, particularly when the countries lack complex refining capabilities. High-value distillates are burned in combined cycle gas turbines (CCGT) during summer electricity demand peaks. Sometimes, distillates displace natural gas completely, when the gas is unavailable or before gas pipelines to power plants are completed.

9.1 The Fuel Mix in MENA Countries: Between Sustainability and Diversification, Options and Consequences

In the Middle East and North Africa, as described earlier, hydrocarbons dominate the electricity fuel mix (steam plants and gas turbines) even if fuel resources vary from one country to another, with natural gas playing an increasingly important role.

Hydro accounts for 6% of the installed capacity. As detailed in the second chapter, there are several factors which contributed to the attractiveness of IPP schemes in the region, one being the very competitive fuel offering (Long-term Fuel Supply Agreements coupled with Long term Power Purchasing Agreements, at below-market prices, usually backed by sovereign guarantees).

In the future, natural gas should remain the fuel of choice, except in selected projects that will be oil-fired mainly because of lack of availability of gas or for strategic reasons.[3] In countries like Saudi Arabia, Kuwait, Oman and Bahrain, where more than half of the gas resources are associated with oil, gas availability for power generation has been problematic. In these cases, it is a significant challenge to account for and then to monetize associated gas—usually flared, recycled or at best reinjected into oilfields. Some of these countries have started to tackle the question and consider how best to use the gas before pursuing gas gathering projects. Others have implemented policies to manage associated gas despite the country's oil-oriented policies. However, most have to face regulatory or technological questions. Given the increasing maturity of some oil producing areas, the greater use of gas for power generation and for reservoir management (reinjection) means it is increasingly important to estimate how much gas is required in upstream applications. Therefore, even the upstream oil sector is gradually competing with the electricity sector for gas supplies.

In such an increasingly strained environment for gas supplies, the authorities in the region are: (a) considering ways to make fuel pricing evolve and (b) thinking about "alternatives" to hydrocarbons. More broadly, the general drive to enhance security of supply translates into several developments.

In fact, energy security has many definitions. The International Energy Agency (IEA) defines it as the 'uninterrupted availability of energy sources at an affordable price'. Energy security has many aspects: long-term energy security deals with timely investments to supply energy while short-term energy security focuses on emergency preparedness and the ability of the system to react to sudden changes in energy supply and demand balances.

For long-term energy security, countries would move to develop a portfolio of alternative fuel sources, even if some bear technological or commercial risks (unconventional resources, imported gas ... etc). Security of supply also entails security of transportation infrastructure. Interconnections are highly relevant, and their reinforcement necessary. Moreover, a robust maintenance is important and strategic repair stocks and spare parts with tower kits are central in emergency repair processes.

The main alternatives for hydrocarbons considered in the region are coal, renewable and nuclear power. Before anything, there are some practical considerations to take into account. First, introducing coal would generally require the development of

[3]Shuaiba, in Saudi Arabia, is the largest oil fired plant worldwide (conventional steam oil 11×400 MW, operational in 2008). It benefited from integrated design and EPC components as well as from lower operating costs.

a coal port given the limited indigenous resources, with relevant facilities that could be shared with other coal consuming industries. These additional costs for import port facilities will have to be taken into account. Moreover, in addition to the reputational aspects linked to carbon emissions issues, oil and gas exporting countries will be careful about the message they would send to their consuming markets should they decide to introduce coal. For these countries, coal for power would have high regret cost if additional cheap fuel supplies become available in the future. For the others, coal is still considered as a viable option in the framework of tight oil and gas markets.[4] Second, burning distillates (usually diesel) to generate electricity is a high cost option but also one of the most flexible since most recent power plants in the region are dual-fuelled and can burn distillates.[5] This is now a widely used practice in the Gulf region during the summer peak period. If pursued on a large scale or long-term basis, logistical, contractual and environmental issues would need to be planned for. It is recommended now that all new GT and CC plants be dual-fired (gas and diesel) and that storage of diesel be provided at each gas-fired station (sufficient for at least 4–5 days of operation at base-load, and at least to run 3–5 days to test the gas supply failure plan). Storage could possibly be enhanced by strategic storage at supply depots and import facilities. In countries with a gas-oriented fuel mix, Steam power plants burning heavy fuel oil or crude could improve fuel security in the event of unavailability of gas supplies but this option is usually weighted against its higher capital cost and lower efficiency. There is limited global experience in operating dual-fired GT and CC plants based on natural gas and HFO, mainly because of the reduced availability of HFO and the relatively high maintenance costs when operating on HFO.

Finally, the "hidden generation" (supply side efficiency) and demand-management are other alternative sources of supply. For example, converting open cycle power plants to combined cycle is a low cost, low regret option that should be pursued in most countries, although it does require re-negotiation of existing contracts. Demand side management (DSM) and energy conservation are also low cost, low regret options, but with institutional challenges due to the large number of entities involved (distribution companies, municipalities, customs department, etc). The cost of DSM is often less than that of new generation, but implementation takes time and results require patience.

[4]Morocco for example still plans to launch its 1320 MW coal IPP at Safi, on a 30 year BOT basis. According the a ONE study, a clean coal power plant will produce electricity at less than $0.05 per kWh, compared with $0.06 for coal gasification, more than $0.06 for wind and more than $0.08 for marine wave energy, and less than $0.04 for nuclear.

[5]In recent power plants, the operator simply needs to push a button to switch from gas to distillates.

Competing Fuels to Hydrocarbons, Assuming the Same Price

For the time being, coal is used only in Israel and Morocco, and there is a remote possibility to see imported coal being introduced in other countries in the region. The current policy of most Middle Eastern and North African countries is generally to use indigenous resources for thermal power generation whenever possible. No change in that policy is expected. The main reasons which would support the introduction of coal are its low cost and its ability to add a substantial generation capacity in a manageable period of time (4–5 years) if pursued on a large scale, and with limited institutional "hassle", compared to other technologies.

On a pure cost basis, coal will have to compete with thermal generation. Figure 9.1 show examples of generation costs in Abu Dhabi and Egypt in 2012. Based on $1–1.5 MMBTU gas price, imported coal can obviously not compete with gas-fired CCGTs, but starts to be competitive if the fuel prices are aligned with their international market value (Heavy Fuel Oil as an example).

To calculate these costs, the following formulas and assumptions were used:

- *Capacity cost ($/MWh) = PMT (Discount Rate, Operating Life, Capital Cost ($/KW))/8.76/capacity factor.*
- O&M cost ($/MWh) = O&M cost ($/KW-yr)/8.76/Capacity factor.
- Fuel costs are calculated depending on the fuel used (gas, heavy fuel oil or coal).
- CO_2 prices are assumed nil. That might have to change in the case of the development of a more comprehensive global framework for carbon trading, or additional development of Clean Development Mechanisms.

With gas prices of less than $1.5 MMbtu, imported coal can not compete with gas fired CCGT, even without accounting for CO_2 emissions. Moreover, regret costs associated with using coal[6] (in case of availability of hydrocarbons' supplies or subsequent discoveries) are not factored in.

Nuclear and renewable are also more expensive than gas-fired CCGT, but they bear other strategic benefits and could act as a hedge against uncertain gas supplies and possible carbon prices or taxes. Also, solar PV costs have dramatically decreased in the last few years.

On the other hand, the two Renewable technologies with natural, economic and technical potential in the region, namely solar PV and wind, still need sometimes a strong and clear supporting policy. Renewable cannot make up a substantial share of the generation fuel mix unless they are backed by a strong governmental will and commitment from the highest authorities as seen in Morocco, and some GCC countries. Concretely, despite large but dispersed solar or wind initiatives in the UAE, Morocco, Egypt and Jordan for example, the contribution by renewable energy sources could remain limited in comparison to the whole generation mix. However, they could be an important policy enabler.

[6]The regret cost here refers to the regret of making a decision to use coal which later turns out to have been unnecessary since a cheaper alternative became available.

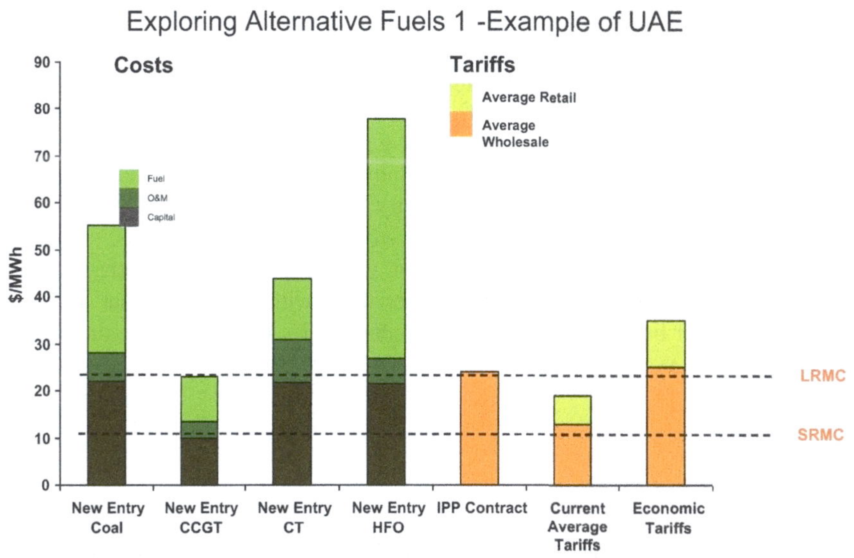

- Power generation costs calculations consider a 10% discount rateand $1.10/MMBTU gas price, which reflects current supplied costs and $4/MMBTU for HFO, $2.75 for imported coal
- Source of IPP - contract estimate from wholesale power purchase price

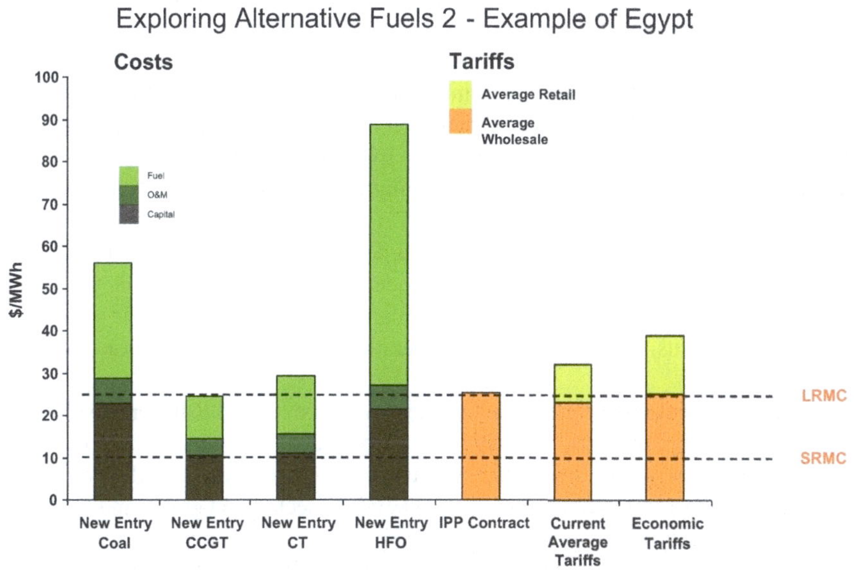

- Power generation costs calculations consider a 10% discount rate, $1.19/MMBTU gas price, which reflects current supplied costs, $5/MMBTU for HFO and $2.75/MMBTU for imported coal
- Current average tariff is the weighted average tariff for all categories of consumers in Egypt. Example of price of PPA:SidiKrir at 2.54 c/kWh – source (http://iis-db.stanford.edu/pubs/20977/Egypt_IPP_Experience_.pdf)

Fig. 9.1 Exploring alternative fuels cost calculation—examples of the UAE and Egypt. Source: Author's calculations

The potential for nuclear power in the region will be discussed in Sect. 9.3. The main conclusions of that section are that its implementation would take more than 10 years, that it would have limited impact on oil and gas supply/demand balances and that there is still a lot to do (institutions, regulatory and educational aspects). Nuclear is in any case viewed as an exclusive prerogative of the State, for strategic, security and geopolitical reasons. It can be introduced without being an impediment to electricity-sector reforms, particularly if it constitutes a limited share of the mix.

As a conclusion, in the long term and with the exception of coal, strong and continuous direct or indirect state support will be needed for other generation technologies to develop further or to materialize, with the caveat that additional cost decreases or breakthrough in storage technologies might expedite the penetration of renewables.

Evolving Fuel Prices

Marginal fuel production costs are increasingly higher than fuel prices agreed under several Fuel Supply Agreements in the region. Energy price reforms are advised to be gradual and sequential. Barring some expectations, fuel price reviews would generally occur before retail electricity tariffs would be increased.

Some countries opt for a testing period of adaptation during which generators are asked to deliver competitive wholesale electricity prices under different fuel pricing schemes. During that period, governments will assess the impact of these new fuel pricing schemes on the cost of electricity, on direct subsidies, and on the attractiveness of IPP schemes in the future (Oman). Evolving fuel prices is an exercise that could be viewed, at first sight, as challenging for the introduction of an additional degree of deregulation.

The need to drive consistent commercial decision-making in the marginal usage or substitution of gas should drive a reconsideration of fuel pricing policy. This can be partially accomplished with notionally higher price assumptions for planning of future investments, in order to impact operational usage. But would actual prices need to change?

Ideally, as mentioned earlier, the first step would be to establish a process to make explicit all energy subsidies, to enable them to be managed with a clear view on their actual size, purpose and their future evolution. More focused subsidy support can then be offered to selected consumers, and key user sectors via grants and tax breaks and/or via structured compensation arrangements, in return for agreed gas or fuel price changes. Examples were mentioned earlier. Then, it would be possible to review and implement the most appropriate approaches to set fuel prices. In the case of gas prices, it is possible to transition to a mix of prices to reflect the balance of pressures. Making energy subsidies more explicit and not applied via energy prices would drive efficiency in energy usage, encourage industrial and other users to seek alternative fuel or feedstock, force consistent and efficient choices in new energy-related investments and respond to real marginal market values. Finally, it will be important to institute a process to continuously review the pricing approach as the balance of the above needs change, or as supply and demand fundamentals respond specific price drivers.

However, re-pricing fuel and gas would impact the competitiveness and potentially the profitability of specific industrial projects and may result in displacing some local production. While a change in gas price would spur some limited, efficiency improvements, particularly in countries with very recent projects, it would also force a "natural selection" between future projects, which in turn would impact the Gross Domestic Product and the macro-economic orientation of the country.

Most countries in the region face today the dilemma of marginal energy costs exceeding long-term contractual legacy supply prices. A number of countries have been in the same situation and have addressed it in a number of ways. They would apply marginal costs to new off-takers only. This is possible when the supply and demand balance is expected to continue to be positive. Others would review long term contracts where possible—using review clauses when they exist in the contracts. Finally, the extreme approach would be a widespread, complete renegotiation, or imposed transition of end-users energy prices to new levels. This last approach may incur breaching long-term supply agreements, and would require a heavy institutional burden.

New fuel pricing can be applied selectively to discriminate between end-use sectors or to distinguish between core use and marginal use. Some countries differentiate between domestic and export industries for pricing (Egypt) or between power and industrial usage. Unification of pricing mechanisms, and use of alternative support mechanisms or subsidies for specific industries, could allow the elimination of distortion of the projects' efficiency and subsequent production decisions.

Options for setting gas prices include cost-related, oil price-linked or linked to freely-traded market prices approaches. Some countries have occasionally linked gas prices to electricity, coal or other commodity price references. The ultimate decision will depend on the relative emphasis needed on driving efficiency of use or on investing in alternative energy sources.

These emphases usually shift over time. Therefore, countries which introduce these mechanisms could adopt a progressive or evolutionary managed approach to re-pricing, i.e. setting prices that are sufficient to drive short-term substitution/efficiency economics without necessarily forcing long term investment solutions which may incur major regret costs. The pricing focus could initially be designed to support efficiency-related decisions—combined cycle, grid reinforcement and integration, efficiency programs in industries. Using the gas supply cost as a reference will be helpful in this regard, without being too punitive for the economy. If more price impact or fuel displacement is needed, pricing could rise to support the next expensive option, such as coal-firing.

This transitioning mechanism is meant to accompany the country as it transforms from a fuel exporter into a fuel "monetizer" before becoming a mature economy (Fig. 9.2).

As mentioned earlier, revising fuel supply agreements will open the question of renegotiating power purchasing agreement terms. More generally, the transitioning of energy prices is likely to require providing some sort of compensation to existing energy users as any material revision of pricing or mechanisms will give rise to

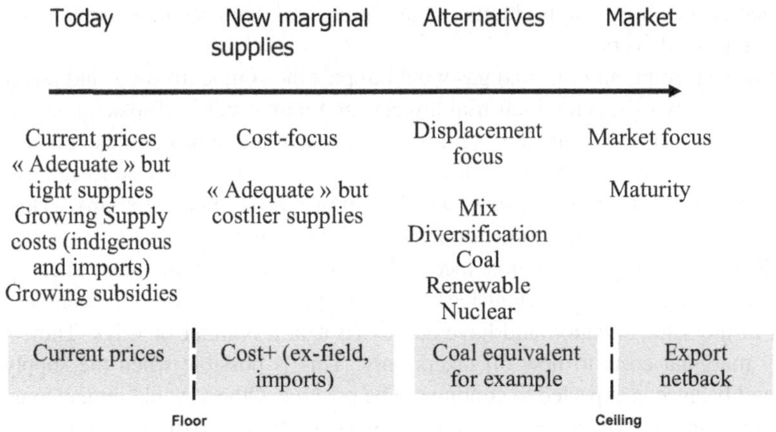

Fig. 9.2 Possible fuel price evolution (illustrative)

claims from contractual gas offtakers. Electricity generation companies might show a certain level of indifference as long as fuel price changes could be passed through. However, beyond a certain threshold, depending on the dispatch rules, affected generators might lose market share to the benefit of lower cost plants.

In theory, the Single Buyer Model is not in contradiction with a fuel diversification policy, however introducing new fuels and new technologies will require the State to step in either directly or indirectly. Regulation will play a critical role to ensure that electricity cost structures and the system efficiency are maintained or enhanced, and that the right investment signals are sent at the right time. Changes in fuel offerings (fuel type, pricing, contract terms...) will require a delicate balancing act. They will test all already established market models in the region and might test decisions to deregulate further, even if they are not a major impediment.

9.2 From Wellhead to Wire: The Position of Natural Gas in the Mix

Barring a few exceptions, most countries with an established gas potential shifted naturally from oil to gas in the 1980s. This followed an international trend in which gas turbines (simple or combined cycle) were considered as the technology of choice for new power plants. More than 90% of installed capacity in Greenfield independent power investments worldwide was gas-fired. Lower capital costs, higher fuel efficiency, lower emissions, shorter lead times were among the reasons explaining the success of gas-fired generation.[7] The introduction of competition in generation

[7]The aviation industry was an important contributer. The technology that made turbines more efficient came from military aircraft and efforts for jet engines' cost reduction was an outcome of the US deregulation of airlines.

supported its growth while capital-intensive coal and nuclear were favoured by vertically integrated monopolies.

In the region, the less gas-rich neighbours preferred to keep a more diversified fuel mix, while securing preferential gas pipeline deals (for example the transit fees in kind between Algeria and each of Morocco and Tunisia; or below-market price deals between Qatar and each of the UAE and Oman through Dolphin).[8] Today, more than half of generated electricity in MENA is gas-fired.

Despite attractive long term fuel supply agreements, the history of investments in the power sector in the region has recorded problems such as occasionally delayed or aborted projects, wrong players attracted by insufficiently high margins, and insufficient investment in transmission. In order to attract more investors, other routes were considered with mitigated success. Some countries tried to develop integrated, power-intensive industrial projects. For example, ALBA (Aluminium Bahrain, 500,000 t/a) has its own 1500 MW power plant and provides up to 275 MW to the Bahraini government during peak periods (summer). Even if the "captive power" scheme brought some additional investments on the supply side, it shows some limitations. It did not improve reserve margins, as it also brought additional demand. By offering subsidized fuel prices, governments set expectations to a new community (energy-intensive industrials) in addition to power utilities. It exposed problems of weak planning and integration into the national system, even resulting in nationwide blackouts (Bahrain).

Another route was to develop integrated gas initiatives, i.e. upstream gas projects that include a downstream component involving power generation and possibly other gas-intensive industries.[9] From a macro economic standpoint (freeing oil for exports, supporting industrial growth) or a more strategic one (ensuring security of supply, rebalancing the power fuel mix), integrated gas initiatives could prove to be successful in meeting the future electricity generation needs as well as developing and monetizing the gas resources of countries with substantial gas reserves. After all, the development of domestic gas resources in the absence of an appropriate

[8]It is becoming more difficult for these countries to secure gas from their resource-rich neighbors. Qatar for example seems to be less in a hurry to conclude additional pipeline exports contracts to gas-tight neighboring countries. It rather prefers LNG exports even for such short distances. There are multiple reasons for this preference. Qatar could play a strong regional role without necessarily being the pipeline gas hub provider. Moreover, regional pipeline gas contracts have been "negotiated" at low levels. Although the "$1 per MMBtu gas" mindset has changed, securing higher prices will be easier through LNG. LNG is flexible and allows for cargo trading and switching (particularly given counter-seasonal regional needs from Qatar's main LNG customers). Most GCC countries have embarked on the same industrialization policy as Qatar using low gas prices as a competitive advantage as well. Pipeline gas projects are at the mercy of political tensions, particularly given the numerous unclear territorial or maritime demarcations in the Gulf region. Even the primary Dolphin concept has partly broken down (prices, costs, nominal "exclusive" supply rights and finally Abu Dhabi developing its own sour gas potential). The project was originally meant to move to further phases and turn Qatar and Abu Dhabi into a regional hub, but this is highly reliant on Qatar increasing its commitments.

[9]One example outside of the region is the Nam Con Son/Phu My gas/power project developed by Petrovietnam and a BP-led consortium.

downstream market appears to be as meaningless as the investment in gas-intensive industries without an appropriate and secure source of upstream supply. Drawing from various experiences in the region (the Saudi Gas Initiative) and elsewhere, there are three key conditions for the success of gas integrated initiatives:

1. Well-defined strategy: the country's priorities and a well-defined gas/power strategy should be clear from the outset.[10] Governments have generally two options: either to be supply-driven or demand-driven. Supply-driven means to put the upstream sector at the centre of the energy policy. In that case, the country would "use" the upstream sector and the integrated gas projects in order to develop the power sector. Demand-driven means that the authorities will prioritize the demand of the domestic market. The implication is usually the creation of a significant domestic market in order to monetize the gas when available. Until the 1980s, natural gas was widely perceived as the weak part of the hydrocarbons chain, usually flared, at best reinjected. These practices continue to prevail in some hydrocarbons provinces. For countries with an export potential, there is obviously one additional question: is it more economically justified to export gas directly for power generation purposes or to generate power domestically and to export it afterwards? In addition to the country's hydrocarbons' development strategy, other determinants include the urgency of the power need, the size of the domestic market and the desired design of the future power sector.
2. Market-oriented offer: The power industry offers lower returns than the upstream sector, making it difficult to attract oil companies. The relatively low rate of return (for IOCs) of less than 9% was one of the reasons behind the shelving of the Saudi Gas Initiative. A critical condition of an integrated project's success and completion lies in a substantial common interest between the involved companies and the host government. A possible solution would be to form upstream-utilities joint ventures, or at least to involve power firms as consultants in a planning and regulatory function.
3. Project scale tailored to the targeted market: Governments face a trade-off. First, there is a contentious issue of security of supply due to the concentration of a significant generating capacity in a limited area. Second, such large-scale projects limit the number of potential investors.[11]

[10]A successful example of gas strategy definition is the Egyptian one as this country clearly emphasized its gas-exporting strategy and its long-term goals.

[11]A good illustrating example of "economies of scale" is the 3000 MW–740,000 t/a Phu My Power-fertilizer complex in Vietnam, which is part of the Nam Con Son gas project developed by Petrovietnam. However, the Phu My project is more the exception than the rule as most integrated gas projects that assumed a large power (and a large power off-take) component remained on the drawing board, e.g. the Southeast Asian Gas pipeline network; and most successfully completed gas initiatives that involved foreign investment actually assumed more modest sizes of power generation capacity building (e.g. The Bilbao terminal, operated by the Bahia de Biskaia (BBG) consortium—Iberdrola, the Basque Energy Agency, Repsol and BP—is integrated with an 800 MW power plant. It is Europe's first integrated regasification and power complex and aimed to link gas reserves in Trinidad to a commercial outlet in Spain).

As stated earlier, natural gas accounts for 67% of the installed capacity in the region and half of the announced projects are gas-fired. At a $1 gas, new entrants can not impose alternative fuels, on a commercial basis only. However, most countries in the region face today the dilemma of marginal energy costs exceeding long-term contractual supply prices in fuel supply agreements to IPPs/IWPPs. A higher gas price will allow for partial switching in the fuel mix. Moreover, the pace of deliverability of gas supplies might not be sufficient to comfortably cover all power generation needs in the region, despite the important gas endowment.

There are 11 countries in MENA with gas reserves of more than 280 Bcm (or 10 Tcf). The total of their reserves constitutes more than 90% of proven reserves in the whole of MENA. 280 BCM is equivalent to the amount of reserves needed for two LNG trains, of 7.8 mt per year capacity each, over 20 years or for nearly five power plants of 1.2 GW capacity each running at 85% of capacity factor during 30 years.

As mentioned earlier, approximately 100 GW were needed to be installed between 2006 and 2015 in the region. If the 100 GW was entirely gas-fired, a 30 year production lifetime at 85% capacity factor would require around 5000 BCM of cumulative gas consumption (and more if part of this gas-fired capacity is open-cycle). That is around a third of the amount produced over 30 years assuming flat production levels. In other words, MENA would need to increase its gas supplies, either endogenous or imports, by more than a third and direct all that output to power generation should it want to have all new projects gas-fired. In fact, since 2000, approximately 40% of new gas production in the region has been directed to LNG exports.

The Middle East and North Africa: Rich Reserves Position, but Capacity Constrained

- Abu Dhabi gas production is flat. The country uses a major part of its gas reserves for injection to assist oil production. Future changes are uncertain, but there are now some new commitments to offshore/sour gas development.
- Qatar projects are subject to a moratorium on new development. Further phases of Dolphin supply: uncertain timing and a vision of "25 MMcfd plateau over 100 years" (around 250 BCM per year).
- Iranian gas development and export projects have been generally frustrated: multiple Pipeline and LNG projects stalled; politics, pricing, alternative gas uses.
- Oman experienced gas tightness. Its future balance is very much dependant on new field developments tendered to International Oil Companies.
- Algeria's gas production is irregular and is constrained by infrastructure issues.

(continued)

- Egypt has been reviewing its gas strategy and pricing given supply tightness and serious shortages during the 2006–2016 period. However, a major offshore discovery (Zohr) is expected to provide some breathing room.

Because of the various issues summarized in the textbox, other solutions were needed to complement the fuel mix. This conclusion will have to be reviewed periodically. The capacity needs of the region were calculated before the economic recession. The 2008–2015 cycle had a double impact. The short-lived slowdown in power demand growth was welcome but a prolonged cycle of lower oil prices might weaken the sustainability of economic development. The risk of power supply constraints in the region is still high, either fuel driven or capacity driven.

9.3 An Opportunity for Nuclear, Amid Proliferation Concerns

Since the beginning of the century, most MENA governments expressed interest in acquiring nuclear technology for civilian purposes, partly spurred by the debate on the Iranian nuclear issue.[12] Egypt, for example, started its program in the 1950s, but reportedly froze it after the Chernobyl accident in 1986. The country has been maintaining a research reactor ever since. In 2007, Cairo announced plans to build a 1 gigawatt (GW) reactor by the middle of the next decade. The Israeli nuclear program also dates back to the 1950s with the creation of the Israel Atomic Energy Commission in 1952 and the construction of the underground Dimona nuclear facility later in the decade. Libya's nuclear history apparently started in the late 1960s parallel to when King Idriss signed the Non Proliferation Treaty in 1968, 1 year before being overthrown by Colonel Qadhafi. There have been several reports on links and activities between Libya and each of China, Russia, Argentina, and Pakistan between the 1970s and the 1990s. These were related to the acquisition of nuclear enrichment technology or the localization of uranium reserves in Africa. In December 2003, after several months of secret discussions between Libya and the United States in the United Kingdom, Tripoli announced the abandonment of its Weapons of Mass Destruction program, clearing the way, at least for a few years, for potential cooperation in the field of peaceful nuclear energy.

Iran's nuclear program has captured the headlines most recently. Its program dates to the era of Shah Pahlavi in the 1960s with the construction of a nuclear research centre.[13] The Shah had ambitious goals for his country and established

[12]Nuclear ambitions are not new to the region. The first efforts date back to the post World War II period, and some continued during the cold war. Some were officially linked to power generation; others were suspected to be for less-peaceful purposes.

[13]The research reactor at Tehran University was reportedly a project under Eisenhower's Atoms for Peace program and was built by American Machine and Foundry Company. It intended for the

enrichment joint ventures with France, open to International Atomic Energy Agency (IAEA inspections), but Iran was then merely an investor with limited access to the technology. After the Iranian revolution, Tehran argued that the Iraq–Iran war merited a resumption of its nuclear drive. The IAEA inspectors discovered a secret deal dating back to 1987 between Iran and Dr. Abdul Kadeer Khan, then head of Pakistan's nuclear program, for the supply of centrifuges and technology for uranium enrichment. In the 1990s, Iran reached out to China and Russia for help. Russia agreed in 1992 to complete the 1 GW Bushehr nuclear reactor. The reactor is nearly finished after a long series of delays and the supply of the fuel had been continuously postponed until December 2007. Discussions have stalled amid the international community's concerns over the real purpose (civilian or military) of the Iranian nuclear program. Meanwhile, Iran made noticeable progress in assembling centrifuges for uranium enrichment and started testing them, in addition to the testing of the machines in its Pilot Fuel Enrichment Plant.[14] The issuance of the National Intelligence Estimate (NIE) report in December 2007 delayed the discussions further. In the meantime, the symbolic connotation and the perceived threat of this progress pushed neighbouring countries to react. They intended to send a message to Iran, to nuclearized countries in the region and to the international community by reverting their past non-nuclear stance. They announced then their intention to revive a dormant program, in the case of Egypt and Turkey, or to develop a new civilian nuclear program, in the case of Jordan, Morocco or the GCC countries. The coordination with the IAEA was stronger than in the past and all declared programs related to civilian applications, including power generation and much-needed water desalination.

First, it appears important to stress that the introduction of nuclear power would not affect oil supply and demand fundamentals. Although the ongoing debate may have an indirect effect on the "psychological" factor driving oil prices, the displaced hydrocarbons would be minimal relative to the global market. A 1 GW nuclear reactor could, in theory, displace a 1 GW crude-fired power plant that consumes around 50,000 barrels per day (bd) of crude or heavy fuel oil. For illustration, Saudi Arabia exports around 7 million barrels per day of crude while Morocco and Jordan typically import 150,000 and 80,000 barrels per day respectively.

---

study of isotopes and other peaceful/medical purposes. Originally, it had highly enriched fuel but apparently in the 1980s was altered to use 20% enriched fuel. The US negotiated intermittently with the Iranian government over a period of about 5 years on a new treaty on the peaceful uses of nuclear power so as to join other countries in constructing power plants under the Shah's nuclear power program. The final accord on the text was reached in 1978, but in view of the political instability at the time was not submitted to the Senate. The revolution froze the talks.

[14]The November 15 2007 report from the IAEA's Director General to the Board of Governors indicates Iran had finished installing 18 164-cascades (around 3000 centrifuges) in Natanz. It reports a level of enrichment of 4%, while Iran has stated it reached 4.8% U-235. The report certifies that the Agency has been able to verify the non-diversion of declared fuel material in Iran, but raises different issues around its own knowledge of the Iranian nuclear program, the non-suspension of enrichment activities, the reactive rather than proactive cooperation, and the non implementation of the Additional Protocol of the NPT.

In countries that have established nuclear infrastructure and institutions, nuclear power generation costs have recently ranged from US$ 55 to 75 per megawatt hour (MWh). Where a particular plant stands within that band depends on a variety of factors. In MENA, below-market domestic fuel prices make it difficult for nuclear power generation to compete on a pure cost-basis with thermal power generation. As these countries transition to higher domestic fuel prices, and as nuclear power infrastructure and regulatory institutions develop, the cost gap between thermal power generation and nuclear power generation would narrow.

However, including nuclear power in a broader power generation diversification effort could bring other benefits. If handled carefully, the ongoing discussions between MENA countries and the international community may lead to the demystification of nuclear power for peaceful purposes. The discussions may help facilitate the debate on other regional geopolitical issues that often fuel the "psychological premium" in oil prices. The engagement of regional parties on the nuclear power debate could contribute to a *détente* in geopolitical problems affecting the region's stability and the anxiety in energy markets. Conversely, a break of the diplomatic dialogue might have a negative effect, spurring covert attitudes and contributing to more geopolitical anxiety.

In the region, the desire for nuclear power is more driven by strategic and geopolitical aspects, transcending pure economics. Given the size of the markets and the increasing importance of the climate change debate, nuclear could play in the future a role (even modest) in the energy mix of these increasingly interconnected power markets. Continuous international and regional cooperation will be critical if nuclear energy for power generation is to succeed. Even if nuclear programs are not new to the region, current discussions around the introduction of nuclear power in the region have gained in complexity, as they come amid the global debate on Iran's nuclear issue and more generally on the application of the Non-Proliferation Treaty and the credibility of international cooperation and institutions. There is a long history of mutual suspicion surrounding nuclear programs. In addition to a comprehensive nuclear legal framework, an ongoing political and economic dialogue with the IAEA will be required, along with technology and fuel suppliers, on access to enriched fuel, safeguards implementation and verification process.

Multilateralism Looked Attractive but Difficult to Follow Through[15]
For political, technological, and economic reasons, working through a regional organization such as the Gulf Cooperation Council appeared as a viable option. Individual initiatives would benefit from the flexibility of bilateral discussions but lack the political "warranty" of a regional organization. The nuclear component of the GCC energy plan, as announced following the 27th Summit of the Supreme Council of the GCC held in December 2006, was endorsed by the Bush Administration.[16] The diplomatic path was closely followed by technical discussions. In

[15]Historical examples include Urenco (a consortium of the British, German, and Dutch governments) and Eurodif (European Gaseous Diffusion Uranium Enrichment Consortium).

[16]The GCC leaders reiterated their call for a peaceful resolution of the Iranian crisis, called on Iran to continue dialogue and cooperation with the IAEA, and to consider the environmental aspects of

February 2007, the GCC signed an agreement with the IAEA to jointly work on a nuclear energy plan, particularly on institutional and other requirements for a joint nuclear power program. In parallel, the GCC + 2 (GCC with Egypt and Jordan) and the Bush Administration have been discussing the potential coordination of efforts to promote regional peace and security, including the nuclear issue. The Saudi Foreign minister mentioned a GCC proposition to establish a uranium enrichment plant in a neutral country outside the Middle East that would serve as a "fuel bank" for Middle Eastern countries, citing Switzerland as an example. Russia, Germany and the United States had already proposed international fuel cycle services in order to guarantee fuel supply, but they made the abandonment of the right to have indigenous enrichment a condition.

In the case of a shared reactor, each country would benefit from economies of scale, assuming smaller units could be developed in the future. By sharing the reactor, neighbouring countries share the output, the risks and the costs (capex, transmission and financing). Nuclear reactors' overnight capital costs could be assumed at around $US 2500–3000 per kilowatt (kW), or a total of $US 2.5–3 billion for a 1 GW reactor. Having a shared nuclear plant would also diminish the demand risk.

In theory, it would take at least 4–5 years, probably longer, to go through the administrative process with the IAEA and to agree on cost-sharing mechanisms, location of the plant, capacity allocations, and other legal agreements. This would be the first regional cooperative agreement to share a nuclear reactor among a group of countries. The only precedent, in Europe, has yet to emerge.[17] In addition to the administrative and negotiations process, another 10 years would be needed to build the nuclear plant. Construction could take around 8 years if there were no delays. In countries with an established nuclear power industry, the "Nuclear Regulatory Commission" requires an additional two or three in order to issue the license to build the nuclear reactor.

There is only a very remote possibility that the international community would enable new nuclear countries in the region to master the whole cycle in house

the issue. They renewed their call for a Middle East free of weapons of mass destruction, commissioned a joint study on the use of nuclear technology for peaceful purposes according to international criteria, and urged Israel to sign the Nuclear Nonproliferation Treaty. When asked about this declaration, The US Department of State stressed that the United States and the GCC have "a very close working relationship and one that has actually become increasingly close over the past months as they perceive the same kind of threats to the region that everybody else has seen." In addition, the US Department of State noted that the United States supports the right of countries to develop peaceful nuclear energy. It also stated that the United States has a great interest in working with individual states that have expressed an interest in developing peaceful nuclear energy. It then cited Egypt as an example.

[17]Lithuania, in the framework of its EU membership negotiations, committed to totally shut down its Ignalina plant by the end of 2009. Given Baltic countries' concerns about gas supplies from Russia and high hydrocarbons prices, they have launched the idea to build a new nuclear plant. A working group was formally set up in July 2007 and started the work on establishing a national investment company with Estonia, Latvia and maybe Poland participating in it.

because of proliferation concerns. In addition, it is uneconomic for an electric utility with a limited number of units to have its own enrichment and other fuel cycle facilities. Countries will still be dependent on a few suppliers to acquire raw uranium ore and enrichment technology. The model Additional Protocol might become the new benchmark for nuclear supply within the Nuclear Suppliers Group (NSG). A regional program would address non-proliferation issues. In such a monitored framework, it appears more difficult to divert fissile fuel or to produce highly enriched uranium for military purposes, more difficult than if every country had to pursue its own nuclear program.

The "Go-it-Alone" Strategy

The six GCC countries are signatories of the NTP. However, the UAE decided to pursue nuclear power alone. It ratified an IAEA safeguards agreement in 2003 and appointed an ambassador to the Agency in 2008. It signed the Additional Protocol in 2009. The country which relied on gas for 99% of its power generation independently published a comprehensive policy on nuclear energy the same year. A regulatory framework and the selection of a site at Barakah followed suit. As recommended by the IAEA, the UAE established the Emirates Nuclear Energy Corporation (ENEC), initially funded with $100 million, and decided to manage its nuclear power program based on contractor services, rather than building up indigenous expertise. The country also decided to forgo domestic enrichment and reprocessing. ENEC sought bids from a shortlist of companies by mid-2009 and appointed CH2M Hill to manage the overall delivery plan. In December 2009 ENEC announced that it had selected the KEPCO-led consortium for four APR-1400 reactors, to be built at one site, for US$ 20.4 billion. The consortium also expects to earn another $20 billion by jointly operating the reactors for 60 years. At the time of writing, the Barakah four-unit nuclear plant was at 75% of construction. There were already concerns expressed by the regulator on the need to scale back the operations of gas fired CCGTs, as the country was originally hoping to export excess electricity to GCC neighbours.

UAE nuclear power reactors under construction and planned

	Type	MWE gross	Construction start	Start up
Barakah 1	APR-1400	1400	July 2012	5/2017
Barakah 2	APR-1400	1400	May 2013	2018
Barakah 3	APR-1400	1400	Sept 2014	2019
Barakah 4	APR-1400	1400	Sept 2015	2020
Total		5600 MWe		

In 2009, the UAE also signed the so-called 123 agreement with the United States, the agreement which would allow the transfer of nuclear energy-related materials and components between the two countries, in return for the UAE renouncing plans to enrich and reprocess uranium or other fuel. The USA will have the right to cancel the agreement if the UAE reneges on its commitment not to engage in enrichment or reprocessing activities.

The construction licence for units 1 and 2 was issued by FANR in mid-July 2012. The 18-month review for the construction licence of units 1 and 2 involved more than 60 staff from the Federal Authority of Nuclear Regulation (FANR), 3 consulting firms, the IAEA. The licences for units 3 and 4 were issued in mid-September 2014.

ENEC awarded six contracts related to the supply of natural uranium concentrates, conversion and enrichment services individually, and the purchase of some enriched uranium product. A spread of suppliers is involved for each stage of the front-end fuel cycle. The company estimates the contracts are worth some $3 billion and will enable the Barakah plant to generate up to 450 billion kilowatt-hours of electricity over a 15-year period from 2017. The country is exploring several options for radioactive waste management strategy, including regional cooperation options.

The UAE's experience confirms that there are multiple important institutional and regulatory milestones needed to pursue a nuclear power program. These include ensuring the existence of licensing and regulation frameworks, process and legal entities, as well as trained personnel to work in these entities and operate the facilities. A comprehensive nuclear legal framework would need to include safety and security aspects, safeguards, and liabilities. Financing for the construction, the maintenance and the safety of the plant, as well as its decommissioning and radioactive waste management would have to be secured. A nuclear regulatory commission would need to be set up to work with the IAEA and domestic players (governments, ministries, utilities), and later to assess design options and evaluate bids. The country will need to prepare a workforce educated in nuclear engineering and operations. In addition to the persons in charge of the bilateral discussions and the personnel employed by the Nuclear Regulatory Commission, one large nuclear plant typically employs 800–1000 workers onsite. Education programs in universities focused on nuclear engineering and technological and education cooperation with selected countries will have to be implemented.

9.4 Green Initiatives: From Oil Exporters to Energy Exporters?

> We agree that we must reduce the environmental footprint of energy production and consumption. But we must be realistic about the cost of green energy. Also about its true potential and how long it will take for commercial-scale supply contributions
>
> ConocoPhillips CEO James Mulva, National Press Club Speech, Washington, January 2009

In the past, clean technologies and CO_2 emissions caps were viewed as a possible threat to the long-term demand for hydrocarbons by some exporters. The global climate change debate was generally snubbed or ignored. At best, energy efficiency featured in some governments' agendas to limit skyrocketing energy demand growth. In parallel, countries with more limited resource endowment in the region have launched renewable electricity projects, with a focus on wind and solar. But in general, there was no major clean energy policy apart from some ambitious targets in

Table 9.1 MENA renewables targets and focus

	Target	Focus	Renewable capacity in 2009 (MW)	Target (MW)
Egypt	20% by 2020	12% wind	305	7200 (2020)
Morocco	10% by 2015	Wind	300	1500 (2015)
Iran	500 MW by 2010	Wind	74	500 (2010)
Tunisia	330 MW by 2010	Wind	20	330 (2010)
Algeria	5% by 2017	Solar Thermal	15	725 (2017)
Israel	20% by 2020	Solar	6	2000 (2020)
Jordan	10% by 2020	Wind	1	600 (2020)
UAE	7% by 2020	Solar	0	1400 (2020)

Source: NREA, IEC, STEG, OAPEC, ONE

the electricity or energy mix. The use of Kyoto Clean Development Mechanism (CDM) Certified Emissions Reductions (CERs) has not been pursued until recently in the region. North African countries, Iran, Israel and Jordan have been the most active in this regard (see Appendix G for the example of Jordan's Renewables policy), spurred by the European Union support and the Barcelona Process (Table 9.1).

Before 2009–2010, the years when solar PV new annual capacity additions accelerated, the definition of realistic targets was important. Government support was central but in this process, external financing and loans have been key as well. However, state dominance would at times discourage investor opportunities, if the political will and regulatory support are weak.[18] These limits are highlighted in cases of large-scale investments like the Desertec scheme, which aimed at implementing

[18]In Shell's long-term "Scramble" scenario, little attention is paid to environmental issues in the early years, with coal being used to plug the gap caused by a lack of easy to develop oil and gas supplies. In the near-term, this would satisfy a "step-change" in energy demand growth caused by the rapidly growing appetite of developing nations. But after 2020, logistical constraints and escalating environmental problems result in a shift in emphasis. Various demand levers must be pulled and new technologies come in, especially second generation biofuels as climate stress becomes apparent. But, even then, neither the demand side measures nor renewable energy supplies are sufficient to avoid surpassing 550 parts per million (ppm) of atmospheric CO_2 by 2050.

To avoid the dangers of the energy security driven Scramble scenario, Shell uses an alternative called "Blueprint" in which incentivized energy efficiency, more renewables and carbon capture and storage are recognized as necessary in the next few years. This is where proactive government intervention comes into play, and the oil industry on its own cannot make the changes that are required. The reduction in energy intensity associated with starting these programs in 2010–2012 should result in lower CO_2 emissions by 2020, making long-term sustainable carbon levels below 550 ppm feasible. Renewable fuels will play a bigger role, but nuclear power will struggle to stay even because of the need to compensate for massive de-commissioning of existing plants over the period. Blueprint scenariois seen as beneficial to Shell as it allows the company to take advantage of its technology edge.

large-scale CSP facilities in the Sahara and connecting them to Southern Europe through High-Voltage Direct-Current (HVDC) transmission lines. The scheme, heavily supported by European partners, lacked the regulatory framework, the infrastructure upgrades and the guarantee that future domestic demand could be covered by other sources.

During that period, Israel was able to push through some initiatives despite a complex institutional and regulatory environment. The country's plans relied quasi-exclusively on private investors (local and foreign) to reach its target. The implementation of the policy was driven by the government, not by the national utility or the ministry, but was constrained by overlapping jurisdictions and contending goals among the various state agencies. Feed-in tariffs of $0.52/MWh for households were agreed in 2008, but large projects still lacked the needed incentives.

Today, the discourse of the hydrocarbons producers in the region has changed, and the speed of renewable development has increased despite the decreasing fossil fuel prices over the last 5 years. Renewables are primarily viewed in the region as fuel savers, with any capacity credits as upsides. The initiatives in various countries are not only driven by a mix of surging domestic energy demand, external support or commercial incentives, as it is too often conventionally stated. The scale of regional investment in renewables and clean energy ventures was around $33 billion between 2009 and 2013.[19]

Moreover, solar PV have been registering record low prices regionally and globally. In 2015, the Dubai Electricity and Water Authority (DEWA) awarded the 200 MW second phase of the Mohammed bin Rashid al-Maktoum solar park to Saudi Arabia's ACWA Power. ACWA had offered the lowest tariff at $59.80/MWh to develop 100 MW under a build-own-operate (BOO) model, and the remaining 800 MW at a lower cost of $54/MWh (lowest cost seen at the time worldwide). ACWA would finance and construct the plant in return for a fixed tariff under a 25-year PPA starting in 2017. DEWA takes a 51% equity stake and provides the land for free. The project is part of the emirate's plans to build 1000 MW of solar capacity by 2030 to meet 5% of its forecast electricity demand from renewables. DEWA is providing the land for free. The low price is driven by several factors, including financing structure of the project, its location (high insolation, and favorable terrain), and the zero-cost land lease. Solar is then already competitive for peak power (against diesel, and even marginal expensive gas). Additional cost declines over the next decade if they materialize, and the scalability of Solar PV, will make it even more competitive. In countries with institutional or regulatory resistance and a dominant government, a hybrid policy framework allowing capacity tendering funded through long term PPAs could enable to increase the role of the private sector.

---

[19]Source IEA. This investment figure is usually compared with the $545 billion earmarked for the hydrocarbons sector during the same period. Although the renewables investments appear to be minimal, it is important to note that several countries in the region are hydrocarbons exporters. Therefore, we argue that the two figures are not comparable.

As far as wind is concerned, recent bids also display competitive prices. Morocco awarded 850 MW of wind at a price of \$31.42/MWh which includes manufacturing of wind towers and blades in the country.

Driven by economic and strategic factors, Saudi Arabia, Abu Dhabi, Algeria or Oman have developed various business models. Some initiatives might be delayed or shelved, but others will survive. That will at least have an impact on the electricity fuel mix, but also on the energy policies of these countries. Solar thermal, more mature than solar PV in terms of deployment, is less demanding in terms of capital costs, but production costs are higher because of the steam turbine process.

Abu Dhabi's \$15 billion Masdar initiative is an example where government's drivers transcend economic and commercial incentives.[20] Masdar was launched in 2007 and included a zero-carbon city, large scale solar projects (the 100 MW IPP Shams 1 followed by Shams 2 and Shams 3), a 500 MW gas fired hydrogen plant, and a pilot carbon dioxide (CO_2) capture project in addition to a CO_2 pipeline network to oilfields (for reinjection). It also has a 20% stake in a wind farm project in the UK and an unspecified equity stake in Finnish wind turbine manufacturer WinWinD Oy. The Masdar initiative is driven by the Abu Dhabi Future Energy Company, wholly owned by the Abu Dhabi government, through the Mubadala Development Company.

Environmental policy considerations became a routine but important part of energy policies in the region. At first sight, the key concern for any energy related environmental policy is in managing emissions of Greenhouse Gases (GHG). Approximately 80% of the region's GHG emissions are attributable to CO_2.[21] However, these levels of emissions remain modest on a world scale.[22] They could grow significantly in the future, particularly as industrial plans are implemented and if coal-fired plants are introduced. Moreover, even if the emissions are relatively small on a global scale, issues of local pollution, air and environment quality are high on the authorities' agendas, particularly in the small GCC countries. This is extremely relevant in countries which register a high carbon footprint per capita like the UAE, where biodiversity is threatened and ecological movements are taking place, generally driven by the country's rulers.[23] The United Nations estimate that Qatar, Kuwait and the UAE have the three highest GHG emissions per-capita, 15 tons more than the United States.

Some countries might push to enhance the measurement, the reporting and the forecasting of CO_2 emissions in the various sectors. But in general, efficiency programs and renewable-oriented policies are relatively easy to implement and

[20]Abu Dhabi's active involvement allowed it to host the headquarters of the International Renewable Energy Agency (IRENA). It pledged to spend \$136 million to fund the Agency.

[21]Source: IEA 2000.

[22]For example, Saudi Arabia emitted in 1998 around 250 million tons of CO_2, and that represented at the time only 1% of global CO_2 emissions and 18% of emissions in the Middle East and North Africa. Source: Earth Trends.

[23]The Emirates Wildlife Society-World Wide Fund for Nature was created in 2001 and is under the patronage of Cheikh Hamdan bin Zayed al Nahyan.

might bear debatable "negative cost" compared with a blanket policy covering CO_2 emissions, while more complex mechanisms like emissions taxes or cap programs might needlessly hamper development programs and divert attention.

Others have introduced incentives for efficiency measures, for renewable energy, incentives for lower carbon power generation, and opportunities for Carbon Capture and Storage (CCS). These have been applied in the oil and gas industry and/or in cooperation with coal-fired power generation, in addition to demand side management and conversion of open cycle power plants to combined cycle.

The upstream sector could play a major role to help introduce some of the mechanisms needed without direct government involvement. The first mechanism would be a more aggressive deployment of solar for distributed generation to cover part of the electricity or energy needs of upstream operations. The second mechanism would relate to the development of Carbon Capture and Storage (CCS) from various carbon emitters (power plants and industries) and inject the CO_2 into oilfields to boost oil production and, in some cases, free up the gas used for reinjection.

In addition to the question over the inclusion in the CDM, there are three other main uncertainties with CCS that make the economics of the schemes unclear. The first one is fundamental and relates obviously to the price of CO_2. The second is the liability period after the injection of CO_2. For example, the EU is debating liability periods of 20–50 years, which are viewed by commercial companies as being too long. Public and political acceptance is a challenge because scientists cannot state with enough confidence whether captured and stored CO_2 will not leak. Finally, costs are still high. The BP-led CO_2 storage scheme in Algeria reinjects CO_2 into the In-Salah field which produces 9 BCM per year of gas. The CO_2 is compressed, dehydrated and reinjected back to the reservoir, resulting in an additional $100 million of investment to reduce the project carbon footprint. Over the project's life, it is expected that about 17 million tons of CO_2 will be stored.

Beyond specific technologies, the institutional feasibility appears important for the development of these green initiatives, whichever form these initiatives will take. The feasibility is reinforced by strong government backing, particularly when funds are available and when the private sector, including in the upstream sector, is willing to share or facilitate sharing technology. In this regard, having centralized policy planning, as opposed to price-driven planning, would act as an enabler. In parallel, facilitating an increased role for the private sector is important. The Single Buyer Model, or an enhanced version, fulfils the two conditions, and yet, it allows the government to keep a major role.

9.5 Summary

Electricity reforms are not in contradiction with a fuel diversification policy, but introducing new fuels and new technologies would require the State to intervene directly or indirectly. This is also relevant for green initiatives and renewable energy, as having a centralized policy planning and a high institutional feasibility would be an enabler. Regulation will play a critical role and changes in fuel offerings (fuel type, pricing, contract terms. . .) will require a delicate balancing act. They will test all already established market models in the region and will challenge decisions to deregulate further.

Out of the 18 countries in MENA, 12 are hydrocarbons exporters (8 of which are OPEC countries). Hydrocarbons dominate the fuel mix, and natural gas, barring a few exceptions, has long been considered as the fuel of choice for power generation. But natural gas will need to be complemented by other fuel sources. In some countries, oil-fired generation continues to prevail because of unavailability of the gas or because of strategic reasons, while other alternatives are considered: renewables, nuclear and coal. Nuclear carries strategic weight but its development in the region would require heavy institutional, regulatory, educational and diplomatic investments and will take a long time to materialize. Evolving fuel prices could be viewed, at first sight, as challenging for any introduction of an additional degree of deregulation, but there are various steps, including a period of assessment and testing that should be followed before tariffs changes. New fuel pricing can be applied to support efficiency-related initiatives, to discriminate between end-use sectors or to encourage fuel displacement. The transition is meant to accompany the country as it transforms from a fuel exporter to a fuel monetizer and then to a mature economy.

Chapter 10
Prospects and Future of Reforms: Resources Monetization

Abstract Natural resources depletion should be offset by manufactured and human capital increase. But there are still wide discrepancies between countries, particularly between North African and Gulf hydrocarbons exporters. Pricing resources below their "social cost" often lead to inefficient consumption, even if this under-pricing contributed to industrial and private sector development. The application of the theoretical framework to two major hydrocarbon exporting countries of the region, Saudi Arabia and Qatar, turns that Saudi Arabia has been building wealth more rapidly than Qatar, a high income country. The two countries seem to be improving the way they manage their economies and their resources.

Keywords Sustainable wealth · Resources monetization · Fuel pricing · Capital

The electricity sector is a vehicle of development and a backbone for economic activity. Its reforms cannot be evaluated in isolation from macro-economic policies and development needs. In order to test the reforms in the framework of governments' general macro-economic policies, another major branch of the economy, the general equilibrium theory and welfare economics, is used.

As per Ken Arrow's definition, the capital assets of a specific country or society include manufactured capital, human capital, the knowledge base and the natural capital. Institutions and capital assets constitute the Productive Base of a country, at any specific time. There is a potential to substitute one category of capital asset with another, but it is difficult to quantify, in the absence of comprehensive and transparent data, the benefits/losses of consuming natural capital for instance. Institutions guide the allocation of capital assets.[1] A "sustainable" growth path would assume that "genuine wealth" at constant accounting prices does not decline over time.[2] The work of Arrow, Dasputga, Ler and others emphasize the need to use other metrics,

---

[1]In Arrow's sense, Institutions include "the legal structure, formal and informal markets, various agencies of government, interpersonal networks, and the rules and norms that guide their behavior".

[2]The concept of sustainability is a complex one, but it became widely used after the publication of the Brundtland Commission report (the World Commission on Environment and Development) in

L. Benali, *Electricity-sector Reforms in the MENA Region*, Perspectives on Development in the Middle East and North Africa (MENA) Region,
https://doi.org/10.1007/978-3-319-96268-9_10

like the Net National Product, rather than GDP or GDP per capita in order to evaluate economic policies and authorities' choices of capital assets' allocation.

10.1 Genuine Wealth Creation and Sustainable Growth

The global quantity of capital assets is assumed limited and finite. Therefore, "genuine wealth" increase in some countries should normally be balanced by a "genuine wealth" decrease in others. Therefore countries' growth paths are not independent from one another.

Olson's work analysed the effects of government and social interactions on economic growth, through models of rational choice based on selfish motives.[3] The vitality of free markets is a necessary but not sufficient condition. In economies with relatively few participants, collective action would be helpful.[4] Olson's analysis is all the more interesting when applied to autocracies: the autocrat, acting entirely in his own interest, might still have an interest in stimulating the economic growth of his country, instead of simply benefiting from rents or abusively collecting others' wealth. This interest would stem from the additional income to tax that will be created by economic growth. Olson develops the notion of an "optimal tax" that would weigh immediate benefits against future better prospects (with the assumption that low taxes benefit economic growth).[5] In this regard, the possibility of growth is greater in the case of royal dynasties (in view of their long time horizon).

Applying Olson's analysis to hydrocarbon-exporting countries in MENA would shed light on some of the discrepancies between the various countries. Failure of growth in North African exporters (Algeria during the pre-2000s, Libya) can be partly explained by the ex-communist countries' experiences. Both countries' socialist systems were inspired by the Russian model:

- Direct State ownership meant "a more efficient extraction of income by the dictatorship", but markets—a necessary condition for growth—did not exist.
- Particular interests had time to organize themselves and remained powerful, even after reforms, particularly as they were able to shape the agenda of these reforms.
- Therefore, the industries and large companies in these countries became powerful stakeholders benefitting from the State's subsidies with no incentives/pressures to be efficient.

1987. Sustainable development was defined as "development that meets the needs of the present without compromising the ability of future generations to meet their own needs".

[3]See Olson, *Power and Prosperity* and *the Rise and Decline of Nations.*

[4]Olson argues that economies with many participants can create the "free rider" problem: individuals have an incentive not to participate but to benefit from the collective actions of others.

[5]Olson applies the same rationale to democracies and concludes that democracies tend to decline over time as particular interests, as opposed to the majority—which will have a stronger focus on growth—would have time to organize themselves.

- Demographics do not necessarily affect economic growth.[6]
- Freedom of property, the rule of law and the sanctity of contracts are important elements that reinforce economic growth. Their absence is disruptive.
- Obviously, the two countries suffered from real disruptive factors related to their own specific recent history (civil war in Algeria and sanctions on Libya). These disruptive events are not completely independent from the structure of the government and the social interactions—assumed to have an effect on growth—in both cases.

Gulf countries, barring variations between countries, appear to have mostly benefited from *"the mutualisation of interests"*. In most cases, there were relatively few market participants and those who existed were directly or indirectly linked to the leadership of the country. Therefore, most had a direct interest in pursuing collective action for economic growth. Family businesses prospered, and even when privatization occurred, genuine markets could emerge—with sometimes mandatory shares of companies offered to the citizens to be traded in the local stock exchanges. Reform or democratization (such as in Kuwait) did not destroy vested interests, like in China. Actually, vested interests in the Gulf had much less time to organize than in North Africa, given their shorter independent political and industrial history. However, the existence of strong economic visions (the UAE), the relative limited number of market participants and the proximity of these participants to the political leadership created a potential fertile ground for economic growth. Increasing hydrocarbons prices facilitated this evolution, with the exception of North African countries which had a different experience.

Measuring countries' performance is not a straightforward exercise.[7] Forecasting it to decide on the best way to allocate resources is even trickier. There is an increasingly wide array of theoretical and empirical literature on the concept of sustainable development and accumulation of wealth and the ways to measure them. These concepts are becoming gradually more comprehensive and more measurable. In a nutshell, the capital assets of a specific country include manufactured capital, human capital and the natural capital.[8] A growing assumption in hydrocarbons producing countries is that there is a potential to substitute one category of capital asset with another. But in general, valuation methods for human and natural capital are still imperfect. Existing economic research falls short of quantifying the losses of natural capital, particularly for hydrocarbons producing countries. However, even these imperfect calculations show that most countries in the Middle East and North Africa failed to make genuine investments in the 1970s and the 1980s when their

[6]Algeria's population is estimated at 39 million inhabitants, Libya's population is around 6 million inhabitants (2013). Both are OPEC countries, and oil and gas producers and exporters.

[7]The UNDP uses today more than 150 composite indices to measure countries' performance, up from less than 25 before 1985. These indices cover a diverse set of issues including competitiveness, governance, social aspects, human rights, environment and security.

[8]More sophisticated calculations include the knowledge base of the society, and sometimes the institutions as assets.

windfall oil incomes could have been used to build up their long-term economic potential. These conclusions, although based on imperfect valuations, caused enough serious concerns among policymakers in some countries in the region to spur them to rethink their resource monetization policy. The concept of Dutch Disease has been widely used to analyse more broadly the inability of resource holders to economically perform in real terms.[9] The empirical research of the low economic performance of hydrocarbons' resource holders was developed during the 1990s. In general, there are two main mechanisms by which the curse would happen: economic distortions linked to the Dutch Disease and/or deficits in governance. The analysis so far has described how some MENA countries attempted to address these two mechanisms, by expanding the non-oil sector of the economy and by rethinking their institutions. The results are mixed, and some countries have performed better than others. The underlying goal here is to revisit the general assumption of low economic performance associated with oil producing countries, in light of global and regional developments.

In theory, the decrease in natural capital should be offset by the increase in produced or manufactured capital and the increase in human capital. Genuine wealth (or a comprehensive measure of wealth) at constant prices should not decline over time to ensure that a specific country is following a sustainable development path, and that it is becoming genuinely wealthier.

This analysis benefitted from the methods developed by Hamilton and World Bank team in recent years to propose comprehensive wealth accounting methods. World Bank (2006) provides assessment of changes in comprehensive wealth in almost every nation of the world. However, several Middle Eastern countries are missing from the assessment. For instance, there is no data for Qatar, and the assessment of Saudi Arabia is incomplete. The work of Arrow et al. (2004, 2008) complements the theoretical framework developed by the World Bank by incorporating technological change and population growth. More importantly, Arrow et al. (2008) offers a theoretical approach to valuing natural resources. Other aspects developed include the utilization of educational attainment instead of simple education expenditure to assess growth in human capital and an improved method to assess environmental damages associated with economic and human activity as they apply to the specific country.

The general rationale of the method is simple, but its application is very much dependant on data availability over a sufficiently long period of time and a deep knowledge of each country's conditions and policies to be able to apply the right metric, particularly for accounting prices. Let V be the productive base and K_t the vector of stocks of all capital assets at date t. V is a function of K_t and if we make the

---

[9]The *Dutch Disease*, the *Curse of Oil*, the *Paradox of Plenty*, among other concepts, have all been extensively analyzed since the 1950s. The expression Dutch Disease was mentioned for the first time in 1977 in "The Economist", in relation with the debate following oil discoveries in the UK. This syndrome refers to the disappointing economic performance of the Netherlands following the development of the Groningue gas field. Empirical research on low economic performance of resource holding countries really took off in the 1990s.

simplifying assumption that t does not directly influence V (which might not be always true in the case of some natural capital), then $V_t = V(K_t)$.

Let K_{it} denote the stock of the ith capital good at date t, and using the chain rule of differentiation:

$$dV/dt = \Sigma\,(\partial V/\partial\, K_{it})\,(\partial K_{it}/\partial t) \tag{10.1}$$

Arrow defines the right-hand side of Eq. (10.1) (*genuine investment or the change in genuine wealth*) as the accounting value of the changes in the stocks of all society's capital assets at t (in other words, the sum of the changes in capital assets' stocks times an accounting price of that capital asset). For example, investment in non-renewable natural resources is necessarily negative. Sustainability requires that real (at constant prices) and comprehensive (including all stocks of capital) capital stock change is increasing:

$$dV/dt > 0 \tag{10.2}$$

While capital stock changes are relatively easy to quantify, the notion of accounting price to apply is more subtle, as it is not typically equal to the market price of the asset. The problem is that the genuine wealth would be highly vulnerable to the chosen accounting price. Arrow et al. (2008) proposes to evaluate the accounting price using the notion of "shadow price". The shadow prices would be "the prices that would prevail if all commodities were traded in commodities' markets and if there were perfect foresight". Applying this definition to hydrocarbons' producing countries would face two hurdles. The first condition is true for part of the oil (exported) and not true for most of the gas exported (LNG long-term contracts). By design, the second condition does not look applicable in empirical analysis, because hydrocarbons' producers can influence future prices while they are not the only ones who can have that influence. Having said that, shadow prices are by nature theoretical prices. For a non-renewable resource like oil, the shadow price would be the discounted value of future use, or the price at which the owner of the well would be indifferent between selling the oil/gas now and holding it for future sale—rental value—or the difference between the sale price of the oil and its extraction cost or the price paid for the scarcity of the resource.

At the same time, when considering a specific country over a long period of time, the rise of the rental value of the non-renewable resource, due to its assumed increased scarcity, needs to be considered. Resource stock owners should expect to receive capital gains and consumers should expect to pay higher prices, or the subsidies' bill of the government to increase if subsidies are maintained.

Extraction costs would be based on the Long Run Marginal Cost of oil or gas supply. These costs would include exploration and development, operations and maintenance expenditures. The Long Run Average Incremental cost would be equal to the discounted value of these costs over the discounted value of the incremental oil/gas output.

The scarcity of resources or depletion could be accounted for by using two methods:

In project finance, a depletion premium is imposed on the economic cost of depletable resources (oil, gas, water ... etc), in addition to their market value or to their marginal extraction cost. It is equivalent to the pre-set value of the opportunity cost of extracting the resource in the future, over its economic price today. Depletion Premium at time t could be defined as:

$$DP_t = (\text{Substitute Price}_T - \text{Extraction Cost}_t) \times (1+r)^t/(1+r)^T$$

R = discount rate and T = time of exhaustion of deposit

In countries where the switching point T is distant into the future, i.e. supply/demand balance in surplus, the present value of depletion premium is negligible and the economic price can be assimilated to the supply cost. In countries where T is imminent or has already occurred, the economic price path will follow the price of the alternative.

There are a few caveats worth highlighting. As mentioned earlier, supply in this section is a direct function of reserves and current output (availability), not of productive capacity (deliverability). Also, this approach depicts the depletion issues at the micro level of a project or of a country. It does not take into account the rarefaction of resources globally and the possible associated 'structural' rise in prices. In parallel, Arrow et al. (2008) decided to consider the capital gains in non-renewable resources as a function of a rate of interest related to the "international capital balance of the resource", by summing up the capital gains of all producing countries. Summing over all consuming countries should be equal to that sum. Therefore, a simplistic way would be to multiply the value of the resource by a "scarcity rent" factor: for example, Arrow et al. (2008) allow the shadow price of oil to increase by 5% per year to account for the scarcity rent.

Similarly, technological changes should further increase the quantity of capital stock. In the case of oil and gas reserves, it is difficult to account for the impact of technological advancements even ex-post, and even in countries where reserves disclosures are transparent. For example, the US Securities and Exchange Commission (SEC) determines the rules by which proved reserves are estimated and disclosed by US-listed companies.[10] The "1978 system", largely unchanged since its creation, falls short of accounting for technological changes and globalization and of "accurately describing industry and individual companies' reserves".[11]

Arrow et al. (2008) offer a more refined approach to account for environmental losses and human capital gains. Instead of using carbon dioxide emissions of a given country as a proxy for environmental damage on its capital, their approach attributes to each country a share of global damages. In this analysis, a CO_2 price is considered

[10] The US share of reserves subject to the regulations was reduced from more than 65% to only 17% of the total as a result of globalization.

[11] For more details on the limits of oil and gas reserves reporting, see CERA's special report of April 2005 "In search of reasonable certainty".

as a proxy for environmental damage in the country and as an opportunity cost of these emissions if they were to be monetized. For the valuation of human capital, they use estimates of average educational attainment. The measure of the aggregate human capital stock increases in line with the average educational attainment and with the population over the selected age cut-off. The price of a unit of human capital would be the rental price for an employed unit of human capital. In the application to the empirical cases of Qatar and Saudi Arabia, investment in education is also used as a proxy, recognizing the limits of this approximation (mismatch between curricula and job market needs, access to education, relevance for high education, etc). MENA countries have yet to address their "monumental waste of human capacity".[12] Some countries are proactively investing in that regard, but it remains to be seen whether funds are being invested in the right place and whether they will yield the desired benefits on development of local human capital in the long term.

Finally, international assets, and liabilities, held by a specific country will need to be factored in. With the rising importance and size of oil producing countries' Sovereign Wealth Funds, this capital share and investments cannot be neglected. In this regard, the 2009 recession resulted in their partial dry-up. The global imbalance, the US external deficit, was cut by half in comparison to 2007, down to $400 billion, mainly at the expense of lower surpluses among oil producers (GCC surpluses fell by around $300 billion during the same period). This picture is mixed though. The SWFs of Qatar, Abu Dhabi and Kuwait, and Libya in North Africa, have been very active abroad during the first half of the recession year, while other SWFs focused on supporting local public projects and banking sectors.

Arrow et al. (2003) calculated Genuine Investment rates for all regions in the world, and concluded that MENA countries were not building wealth (−7.1% over 1975–2001). The same team offered in 2008 a more nuanced approach explaining that they could have understated sustainability of Middle Eastern development growth. A more refined approach applied to the first years of this century is offered. For the time being, preliminary estimates for some gulf countries point to the fact that this balance is improving (see Sect. 10.3 for examples of Qatar and Saudi Arabia). However, a refinement of the calculation will be needed in order to develop a more comprehensive view. It is far from certain that the development growth path undertaken by gulf countries today is optimal, but it appears more sustainable than economic policies followed in past periods of high oil prices. The central question is

[12]As referred to in the first chapter, education and employment are becoming high on Arab countries' agenda, as they recognize the vulnerability of their economies because of the region's poor track record on education, training and human development. "The Arab world's track record on education, particularly girls' education, is discouraging. Sixty-five million adult Arabs are illiterate and two-thirds of them are women. More than 10 million Arab children between the ages of 6 and 15 are still not enrolled in any schooling, and on current trends this number will increase by 40% over the next decade. This is a monumental waste of human capacity." H.H. Sheikh Mohammed bin Rashid al Maktoum, *Education vs. Extremism*, the Wall Street Journal, June 3 2009.

whether their hydrocarbons resources are *properly valued* to ensure genuine wealth creation and sustainable growth.

10.2 Fuel Price Changes: Testing the Long-term Fuel Supply Agreement Paradigm

Hydrocarbons are direct or indirect inputs to the production of consumption goods. Authorities in the region tend to price exportable natural resources well below market price and electricity services below cost of supply.

As mentioned earlier, electricity generators in the gulf countries have been receiving a below-market gas price of around $1–1.5 per MMbtu. Other fuels, such as crude oil and Heavy Fuel Oil have also been priced at below-market prices. Similarly, subsidized gas prices enabled these same countries to attract large electricity intensive industries (aluminium) and gas-fed industries (petrochemicals) and to support industries using natural gas as a fuel in their process (such as cement).[13] Electricity is in turn sold at subsidized tariffs. Some industrial products have been marketed at regulated capped prices in the local market (e.g. cement in the UAE).

It is generally accepted that pricing consumption goods below their "social cost" often leads to excessive, and therefore inefficient, consumption. This excess of consumption, beyond a level of "genuine consumption", would not contribute in any way to economic growth. On the contrary, it will accelerate the depletion of the natural capital of the country without adding to the genuine wealth. On the other hand, it could be argued that without this under-pricing, industries would not have developed at the same pace and private spending would not have increased at the same rate. This argument associates under-pricing to tax cuts that would stimulate private spending and investments. Therefore, this hidden benefit (positive externality) should be included in the social cost. The reasoning can be pushed even further in case of the creation of further downstream industries, which could not have existed without the establishment of the primary industry (in the case of aluminium for example). The multiplier effect (on jobs, on economic growth, etc) of these downstream industries can be seen to offset the loss in efficiency that would result from an under-pricing of fuel or energy supply to the primary industry.

The structure of the upstream sector (one monopolistic National Oil Company and/or multiple upstream players) is central. In the case of hydrocarbons producing countries with a large enough pool of private upstream players (such as in Egypt), the upstream companies would at least adopt a cost-plus approach, to ensure the

[13]Fuel supply agreements for industrials are generally highly confidential. In the region a few fell in the public domain, and they are also around $1 per MMBtu in the case of gas, sometimes with escalation formulas.

viability of the project. They would consider the opportunity cost of the fuel they produce and sell to the authorities, instead of exporting it if they were allowed to do so. However, if part or all of the production has to be sold domestically, the minimum acceptable price would need to cover the cost of the development and operations. For example, in Egypt, upstream operators were able to secure higher prices from the government since early 2008 to reflect higher operations costs.[14] In this context, the involvement of private stakeholders in the chain means that fuel pricing is continuously reviewed and challenged as the supply mix evolves. When a single National Oil Company operates in the country, the relationship with the authorities is more complex, although it should lead to the same outcome in the long-term. Indeed, National Oil Companies are expected to play multiple roles in addition to primary role of fuel supplier through their upstream and downstream responsibilities. They are generally expected to contribute to the welfare of the citizens.[15] However, as companies become corporatized and increasingly questioned on their capacity to deliver, the argument is usually made that fuel prices should enable them to cover their costs, to protect their financial sustainability and their capacity to invest in a highly capital-intensive industry (see textbox on NOCs' funding).

National Oil Companies' Funding

Algeria's Sonatrach competes with the government for oil revenues, in the absence of a clear fiscal and dividend regime. Between 1980 and 1994, when foreign involvement was virtually inexistent, oil revenues were split with, on average, 67% in favor of the government. In crisis periods, such as 1994, the national company had to pay additional levy on its profits. Sonatrach is also deprived of hard currency, since exports revenues are immediately transferred to the Central Bank, before being shifted as bank deposit in the local currency.

Saudi Aramco pays taxes and royalties to the state and dividends to its sole shareholder (the Saudi state). The company enjoys some freedom in setting up its budget and engaging expenses, under the monitoring of the Ministry and of the Supreme Petroleum Council (now replaced with the Supreme Council for Saudi Aramco).

The National Iranian Oil Company (NIOC) had no control on its revenues. The proceeds of oil sales were immediately transferred to the Treasury

(continued)

[14]In early 2008, the Egyptian cabinet adopted a new maximum price of US$3.95 per MMBtu for the gas produced in five different concessions in the Mediterranean Sea (the BG-operated Rosetta and West Delta Deep Marine (WDDM) concessions, RWE's North Idku, and Eni's and Hess's North Bardawil and West Mediterranean Block 1 concessions). It represented a sharp rise from the previous maximum price of US$2.65 per MMBtu, but is in line with the 2007's bilateral deal between the Egyptian government and BP, giving the company a US$4.5 per MMBtu price for its gas to compensate for high exploration and development costs in recently discovered fields.

[15]See Marcel, *Oil titans*.

and NIOC was allocated a budget within the State's budget by the Parliament. The 2006 budget law allowed NIOC to retain 7.3% of oil sales before transferring the remainder to the Treasury. NIOC keeps the proceeds of sales of condensates and gas.

Egypt Gas and Petroleum Company (EGPC) has been paying to the Government a royalty of 10% of the hydrocarbons produced, cash or in kind. It contributes also partly to the subsidizing of gas and products in the domestic market by "small" transfers to the Government's budget.

Qatar Petroleum's oil and gas activities are the primary source of revenue of the national budget. The company pays to the Ministry of Finance royalties and tax revenues on export sales of crude, refined products and gas products. The State also receives dividends from QP's affiliated companies and other businesses where it has direct and/or indirect stakes.

This is all the more relevant now that fuel supplies are becoming costlier to develop and to produce. A few reasons include the maturity of upstream assets, higher energy intensity of upstream production, increase of EPC costs due to tight service sector markets, undeveloped resources in increasingly more complex structures, tight reservoirs, or containing impurities (e.g. sour gas).[16]

At the time of this writing, all countries are at different stages of reviewing their fuel pricing structure. This reconsideration is sometimes spurred by the upstream players, or by the scarcity of supplies or both. A few have already renegotiated their long-term Fuel Supply Agreements (FSAs).

For utilities, FSAs' amendment often leads to a revision to the Power & Water Purchasing agreement (PWPA). This could indicate that the electricity generator will renegotiate the price at which its sells its output to the Single Buyer. If the fuel is priced at its real cost, governments will simply be transforming the double subsidy applied to the electricity sector into a single direct one to the distribution companies, as explained earlier. Utilities would also take advantage of this situation to renegotiate other features of the contract, such as non-dispatchable provisions.

Even if renegotiation of contracts usually sends negative signals, this does not necessarily imply that international utilities would be less inclined to bid for IPPs in the region if gas prices are higher, provided they can pass through these higher fuel prices to the offtaker, i.e. that their output selling price is also higher and guaranteed.

In a nutshell, in the MENA region, fuel dynamics were a major factor behind renegotiations of long-term agreements, and not electric reforms. The changing forces at work and the structure of the upstream industry (private players, scarcity of supplies, increasing costs, evolution of international oil and gas prices, to cite a few) were key drivers.

---

[16]The cost of the Saudi marginal barrel of oil is estimated at around $20 per barrel (source CERA). New gas production cost (sour gas in Abu Dhabi for instance) is in the area of $5 per MMbtu.

In the case of industrials, particularly those exporting their output and building their competitiveness and margins on the low-priced and guaranteed fuel supply, this process is much more complex as it is more difficult for the fuel regulator (ministry or another institutional regulator) to access timely information on the actual profitability of these industries. Negotiations are seldom driven by higher fuel supply costs, but rather only when there is an issue with the availability of the fuel, or even a Force Majeure clause. In some cases, a preliminary "grand bargain" would take place beforehand with attempts to evaluate resources utilization and monetization. Industries' contribution to the economy and to the society would be scrutinized. In general, industrials want stability of supply and guaranteed prices, preferably via long-term contracts or via the construction of their own captive plants. They would prefer to either sign long-term guaranteed supply contracts, or to build their own captive plant, with the possibility to sell the surplus to the grid if they are allowed to. Ideally, this complex evaluation should take place before launching any industrialization program, or at least before supplies become tight, and should be reviewed periodically. However, there are several examples in the MENA region and in the rest of the world of countries who overcommitted their fuel resources at below-market prices, and faced a supply crunch because the "grand bargain" did not happen on time, or was launched when it was too late.

From the government's perspective, tight supplies would induce "discrimination" between projects and sectors, first on the basis of fuel substitution potential. In this regard, differences between gas-fed and gas-fired industries can be taken into account. Often, the country would be committed to serve all domestic demand, i.e. at the expense of exports, even when this demand is not justified on an economic and financial basis. Ultimately, the electricity sector has a near absolute priority ("we should keep the lights on"), even when it could be the most flexible sector in terms of fuel switching.

10.3 Monetizing the Resources: Empirical Cases of Saudi Arabia and Qatar

Opting for the examples of Saudi Arabia and Qatar as case studies for the multifaceted subject of resources monetization is not accidental. Although very different, the two are major oil or gas gulf exporters who have prudently reformed their electricity sectors. Both have embarked on a first wave of ambitious industrial programs based on below-market priced fuel supplies. Yet, their development growth path appears very different. To enable a comprehensive comparison, this section will focus on the macroeconomic fundamentals of the two countries during the first decade of this century, inclusive of the investment choices and policies during the 2008–2009 global financial crisis. This will shed light on the macroeconomic and political orientations of the two countries during the 2010–2020 decade.

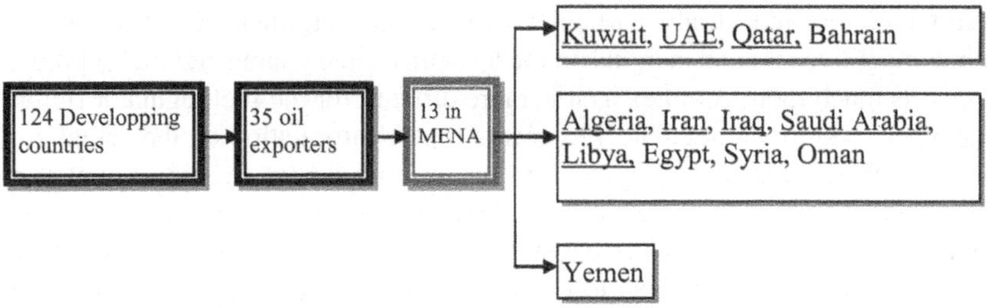

Fig. 10.1 Oil exporting MENA countries World Bank classification

The World Bank classifies Saudi Arabia as middle income and Qatar as high income countries. Is this classification correct? In 2008, Saudi Arabia's GDP per capita was $19,405 while Qatar's was $106,000.[17] Several factors, including the geographic and demographic size of each country, predetermines the macroeconomic potential of a country, however the key question of whether that potential can be reached remains (Fig. 10.1).[18]

The Saudi 2009 budget points to a −4.6% of GDP deficit, its first deficit since 2001. The country decided to run on an expansionary budget, despite the relative fall in oil prices.[19] In comparison with 2001, domestic debt has dramatically decreased from nearly 100% of GDP to less than 10% in 2008. Over the same time period, government's deposits with the banking system increased from 7 to around 70% of GDP, while SAMA's foreign assets grew from less than 25 to nearly 100% of GDP. The non-oil fiscal deficit has however expanded from around 20% of the nonoil GDP to more that 60%. This highlights the growing importance of government's spending in the Saudi economy. The 2009 budget is particularly instructive as the increase in capital expenditures is in marked contrast with previous periods of decreasing oil prices. Capital spending in the 1998 budget, when oil revenue dropped by 50%, declined by 15%.

Building up on the 1970–1974 plan diversification imperatives, and the strong growth in construction and financial services witnessed in the 1970s (10.3% and 15.0% growth, respectively, during 1970–1980), a massive industrial and infrastructure expansion occurred in Saudi Arabia after 2003. In addition to the impressive increases in public spending, supported by increasing oil prices, the energized private sector, supported by government spending, has been channelling investments

[17]Source: OPEC annual Statistical Bulletin. Qatar came second in the world after Liechtenstein, straddled by Austria, Germany and Switzerland.

[18]Saudi Arabia's population was around 24 million in 2008, while Qatar accounted about 1.64 million inhabitants. The former extends over an area of more than 2,240,000 km^2, whereas Qatar registers 11,400 km^2.

[19]The oil price assumption in the budget is $37 per barrel.

into several economic sectors. The utilities sector benefitted from this momentum. Albeit slow, this shift occurred under the stewardship of the Saudi leadership.[20]

In terms of human capital, unemployment is a major issue in the country and continues to pose a challenge. Latest official estimates put joblessness among nationals at 12% (and higher among young people—probably 25% for the 25–29 age group). The authorities have invested heavily in vocational training, and have tried to address the mismatch between school curricula and job market needs (more technical and scientific orientations).

As far as natural capital is concerned, the focus here will be on hydrocarbons. Saudi Aramco conducted a major investment program to increase productive capacity to 12.5 million barrels per day (mbd) by 2009. In 2012, it announced that to maintain the capacity at 12.5 mbd it would spend $35 billion over the following 5 years. Gas exploration and production has been a major part of the company's capital program.

As far as foreign assets are concerned, their value depends on the investment strategies and on the international markets' dynamics given that most of these assets are invested internationally. For example, Saudi's investments in foreign securities and foreign banks plunged by a third between December 2008 and June 2009. This was coupled with a massive withdrawal of funds from international markets, because of the expansionary budget of the country and lower oil prices.[21] In parallel, the country's public debt has decreased to 13% of GDP at the end of 2008, from more than 100% in 1999.

Qatar registers the second highest GDP per capita in the world, but the Qatari authorities were increasingly conscious that this criterion is far from being sufficient to ensure accumulation of wealth, genuine investments and the best use of natural resources for current and future generations. In other words, the question is how to ensure that Qatar's genuine wealth per capita at constant accounting prices does not decline.

[20]Cordesman (2003) writes: "Fortunately, most informed Saudis—including the Crown Prince and the vast majority of senior princes, technocrats and businessmen—understand that Saudi Arabia must make major structural economic reforms. These "modernists" are as Islamic as every other element of Saudi society, but they are supported by most well-educated Saudis and a generational shift in favor of reform. They realize that Saudi Arabia must cap or reduce its funding of social services and must focus on a pattern of economic development that will ensure that all national groups receive a share of the nation's oil wealth. They recognize that reform must take place within a safety net that defuses or reduces social protest. At the same time, they understand that oil wealth is as relative as any form of wealth and that it is steadily declining in terms of per capita income. They see the need to diversify and privatize the economy and to create a competitive economic environment dependent on Saudi labor".

[21]"Saudi Arabia is using its overseas assets to finance domestic requirements specified in its budget for this year ... in other words, the kingdom is channeling part of those funds into domestic development because it apparently does not want to borrow at this stage," Malick Yunus, senior economist at the National Commercial Bank, July 2009.

The country was able to build a strong economic presence in the region and internationally. The "healthy deficit" (2% of GDP) earmarked in the 2009–2010 budget is a signal that the authorities will continue to stimulate the economy by maintaining a high level of state spending and to maintain high growth rates, despite the global economic turndown and the fall in hydrocarbon revenue. LNG exports are designed to complement more volatile oil export revenue. The high GDP growth rate of more than 10% registered in the early 2000s, in real terms, declined to still strong rates north of 6%. The external debt has been maintained at 60–70% of GDP, and the fiscal balance in constant surplus. Official reserves have increased to $15 billion in May 2009, up from $9.8 billion in 2008 and less than $2 billion in 2002.[22] They stood at $43 billion in December 2014 (Qatar National Bank). Qatar's sovereign wealth fund Qatar Investment Authority (QIA), set up in 2005, held $60 billion in assets in 2008. QIA has faced difficulties in 2009 because of its investments in Barclays Bank and in western real estate, but it was critical in bailing out parts of the Qatari economy.[23] The 2008 global recession hit some hydrocarbons and petrochemicals projects, in addition to tourism, real estate and retail, but the country is well perceived and still able to tap the bond markets. Inflation moderated to 1.3% in the first quarter of 2009, down from its 17% peak reached in the first half of 2008.

Qatar's macro-economic policy has been anchored on the objectives of the "Qatar National vision 2030", for human, social, economic and environmental development. The key component of this vision is to allocate a large part of government's spending to "strategic" and infrastructure projects. The Emirate had bet on a rapid expansion of LNG production and investments aimed at economic diversification. But the country has three main economic dilemmas. First, Qatar is a relatively small country and there is a limit to what its system can sustain.[24] Overinvestment and overconsumption, particularly in the industrial areas, translated into bottlenecks and logistical constraints. Second, it witnessed double-digit inflation rates during the 2000–2010 economic boom.[25] Finally, the authorities would like to see most of their income generated from non-energy related industries by 2020. Given the emphasis on energy-intensive industries, "real" diversification appears difficult to achieve.

The first two dilemmas were momentarily and conveniently solved by the global economic crisis. Several projects were delayed providing some breathing room, although major ones still went ahead.[26] The inflation peak of 16.6% reached in the second quarter of 2008 eased down to 8–10% in 2009.[27] However, economic

---

[22]Source: Central Bank and SAMBA.

[23]As of mid-2009, QIA was set to buy 10–20% of the capital of listed banks ($5.3 billion), in parallel to a $1.8 billion plan to buy banks' investment portfolios and a $4 billion plan to acquire banks' real estate assets.

[24]11,400 km^2, 1.64 million inhabitants (Official figures for mid-2009 report 1.64 million inhabitants, including expatriates, double of 2004).

[25]Inflation reached 11.8% in 2006, 13.8 in 2007 and 15% in 2008, according to the IMF.

[26]The MEED projects tracker estimates that the value of projects under construction and in the planning stages in 2009 stands at $200 billion.

[27]Source: IMF, National Bank of Kuwait, Standard Chartered.

diversification can not be solved "coincidently". The Comprehensive Industrial Development Strategy of Qatar relied on vague and complex concepts. It was deemed as not comprehensive enough and not easily implementable.[28] The dilemma posed by economic diversification, away from hydrocarbon and hydrocarbon-based industries, was probably too difficult to solve. Qatar encouraged the creation of small and medium sized industries, particularly in industrial estates and cities. But it was equally conscious to take advantage of its resource endowment and develop larger energy-intensive industries as a way to invest its resources locally.

As far as electricity reform is concerned, and like most other countries in the region, Qatar introduced competition in the generation of electricity and implemented a single buyer model. In 1990, it established the Qatar Electricity and Water Company (QWEC) as a joint stock company with government as major shareholder, and started transferring to it the assets of the ailing Ministry of Electricity and Water in 1998. The Qatar General Electricity and Water Corporation (Kahramaa), created in 2000, is the sole buyer of electricity and water and has signed long-term power and water purchase agreements with independent water and power producers (typically 25 years). QWEC, now 57.2% privately held, had all government-owned generating assets gradually transferred to it, with the view of future privatization.[29] QWEC is the shareholding partner in projects with international partners. AES won Qatar's first IWPP in 2001 (Ras Laffan A).

The aim was to "complete" the deregulation and privatization process in mid-2002. The government has had plans to privatize and to increase water and electricity tariffs for foreigners and commercial users over a 3- to 4-year period. It even hinted at charging the Qataris for electricity if electricity demand growth does not decelerate. Power and water are free for Qataris.[30] Today, as far as power tariff setting is concerned, cross-subsidies apparently exist. Other customer classes are subsidizing, for example, relatively low industrial tariffs ($0.019 per kilowatt-hour [kWh]). Qatar increased electricity tariffs in 2002, and then introduced a new hike to electricity and water charges in September 2015. Industrial tariffs were increased to QAR 0.13 per kWh ($0.035 per kWh).

Even if it was not the first Middle East country to launch an electric reform, Qatar is the first one to have nearly fully privatized its power and water sector, even if the government keeps a large share of QWEC (direct holding of 42%).

---

[28]In this respect, the objectives of the Strategy emphasize concepts like "the maximization of value addition from raw and intermediate materials, the diversification of income resources, the increase of manufacturing industries' contribution to GNP, the industrial awareness for a sustainable industrial development and self-reliance, the scientific and technological developments and the protection of global environment".

[29]Ras Abu Fontas A & B, Ras Abu Abboud, and three satellite stations: al Wajbah, Saliyah, and Doha South. This asset allocation reminds of another similar move in Abu Dhabi with TAQA (which has gradually invested in international assets).

[30]Electricity was free for Qataris up to a monthly ceiling. Expatriates paid between $0.021 per kWh and $0.027 per kWh and commercial customers up to $0.037 per kWh.

Table 10.1 Saudi and Qatar oil prices

In $ per barrel	2000	2001	2002	2003	2004	2005	2006	2007	2008
WTI	30.36	25.92	26.15	31.05	41.49	56.59	65.89	72.22	99.72
Saudi Arabia									
Arab Light (32 API, 1.8% S)	26.75	23.1	24.2	27.7	35.6	50.2	65.9	68.7	94.7
Arab Medium (31 API, 2.5%)	25.98	22.6	23.8	25.2	33.5	47.7	63.8	66.4	92.3
Arab Heavy (27 API, 2.8%)	25.19	21.9	23.3	24.5	31.7	45.5	62.8	64.6	90.1
Qatar									
Dukhan (40.9 API, 1.27%)	26.98	23.7	24.3	29.1	35.9	53.2	66.4	71.5	97.9
	2000	2001	2002	2003	2004	2005	2006	2007	2008
Saudi Arabia									
Arab Light	65%	65%	65%	65%	65%	65%	60%	60%	60%
Arab Medium	13%	13%	13%	13%	13%	13%	15%	15%	15%
Arab Heavy	22%	22%	22%	22%	22%	22%	25%	25%	25%
Qatar									
Dukhan	100%	100%	100%	100%	100%	100%	100%	100%	100%
Weighted oil prices									
Saudi Arabia	**26.31**	**22.79**	**23.97**	**26.69**	**34.51**	**48.85**	**64.82**	**67.31**	**93.20**
Qatar	**26.98**	**23.74**	**24.30**	**29.05**	**35.91**	**53.21**	**66.42**	**71.48**	**97.92**

Source: OPEC, Platt's
Bold values are the results of the weighed average calculation

Genuine Wealth Creation in a Period of Increasing Oil Prices

In order to assess whether Saudi Arabia and Qatar have been building wealth during the 2000–2008 period (increasing oil prices), formula 10.1 as described in Sect. 10.3 could be applied. Given the low extraction costs, export oil prices for each of the main crudes exported by the two countries could be used (Table 10.1):

As far as gas is concerned, for simplification, the maximum export netback price of QatarGas IV could be used as a proxy for gas price valuation. As a reminder, the shadow prices would be "the prices that would prevail if all commodities were traded in commodities' markets and if there were perfect foresight". Therefore, the export netback value could be used even in the case of Saudi Arabia which does not export gas and uses it exclusively for domestic consumption. Natural gas used for reinjection would be considered mostly recoverable, barring a loss of 5–10% (Table 10.2).

Tables 10.3 and 10.4 summarize the results for the two countries. There are several caveats. Data generally differ between various official sources, including ministries and bear many inconsistencies. The least probable data are highlighted. Also, CO_2 was priced in this exercise at $20 per ton, considered as the opportunity cost of these emissions. Another caveat is that the financial investments and the role of the increasingly sophisticated Sovereign Wealth Funds and other investment

Table 10.2 Historical export netback prices for Qatar FOB

Date	United States	Europe	Japan	Maximum
2000	1.81	1.03	**3.01**	3.01
2001	1.51	1.79	**2.91**	2.91
2002	0.87	1.33	**2.59**	2.59
2003	2.98	2.18	**3.08**	3.08
2004	3.37	2.42	**3.48**	3.48
2005	**6.32**	3.95	4.29	6.32
2006	4.27	**5.96**	5.39	5.96
2007	4.47	**6.11**	6.01	6.11
2008	6.37	9.72	**10.48**	10.48

In $ per MMBtu FOB

Assumptions: Liquefaction costs at $1.53 per MMbtu and Regasification at $0.17 per MMBtu. Shipping depends on destination, for example, $1.1 per MMbtu from Qatar to Japan
Source: Cedigas, IHS CERA
Bold values correspond to the maximum value

vehicles are only partly taken into account. The assessment of assets in foreign countries is based on publicly available information and is therefore limited. Likewise, education expenditures are limited to governments' direct expenditures in fundamental education. They do not include professional training or human development and do not capture inappropriate curricula (lack of scientific or engineering orientations). Finally, and more specifically, the population in Qatar has doubled since 2004. This 'artificial' increase is due to the intense economic activity and the immigration of expatriates, usually for short-term periods. Therefore, for this calculation, the "natural" growth of the pre-2002 population in Qatar was considered.

At first sight, it appears surprising, given the differences in GDP per capita, that Saudi Arabia's perspectives look brighter than Qatar (genuine investment rate of 9.4% for Saudi Arabia and −4.6% for Qatar), but as stated earlier, financial investments were only partially included in the calculation and that might be penalizing Qatar. Moreover, Qatar is highly vulnerable to the shadow price of natural gas, more than Saudi Arabia is vulnerable to oil price.

The maximum export netback price of LNG might be overstating the depletion of Qatar's gas reserves, as a large share of LNG exports are bound in long-term contracts. Table 10.5 illustrates the same calculation but using the average netback price to all three markets instead of the maximum. In this case, the balance becomes positive.

For these two countries, during the first decade of the twenty-first century, the balance offsetting natural resources decline with human and manufactured capital has been improving. For all the uncertainties mentioned above, the two countries are improving the way they manage their economies. If the structural price of hydrocarbons increases in the future, the energy depletion element will become more important. It will be even more critical to offset it with accelerated investments in human capital, in industries and in infrastructure.

Table 10.3 Genuine wealth creation in Saudi Arabia

Saudi Arabia	2000	2001	2002	2003	2004	2005	2006	2007	2008	2000–2008
GDP ($ millions)	188,440	186,240	188,530	211,200	212,000	*270,000*	353,000	378,000	481,000	2,468,410
Population (millions)	20	21	21	22	23	23	24	24	25	
Foreign assets ($ millions)										59,000
Domestic Net Investment ($ millions)[a]	4897	8435	8000	8925	10,014	16,614	18,910	31,773	*40,000*	147,567
Education expenditure ($ millions)	*13,200*	*14,200*	*14,500*	*15,300*	*17,000*	*18,000*	*20,000*	25,000	28,000	165,200
Damage from CO_2 emissions ($ millions)[b]	−5160	−5300	−5500	−5780	−6080	−6480	−6740	−6942	−7150	−55,133
Energy depletion ($ millions)	−3674	−3764	−3345	−4585	−5978	−12,602	−12,024	−12,591	−26,497	−85,061
Genuine investment rate (%)										**9.4**
CO_2 emissions (Mt of CO_2)[c]	258	*265*	*275*	289	304	324	337	*347*	*358*	
Crude oil production (in mbd)	**8.00**	**7.80**	**7.09**	**8.40**	**8.90**	**9.35**	**9.20**	**8.80**	**9.2**	
Gas production (in bcm)	**49.8**	**53.7**	**57.3**	**60.1**	**65.7**	**71.2**	**73.5**	**74.4**	**80.4**	

Sources: IMF, SAMBA, JADWA, OPEC, IEA

In yellow, forecasts, estimates, budgets and author's assumptions

[a]Does not take into account financial investments

[b]CO_2 cost assumed at $20 per ton

[c]IEA, and author's own estimates after 1999

Table 10.4 Genuine wealth creation in Qatar

Qatar	2000	2001	2002	2003	2004	2005	2006	2007	2008	2000–2008
GDP ($ millions)	*16,000*	*19,000*	*20,000*	*25,000*	*35,000*	43,000	57,000	*71,000*	91,000	377,000
Population (millions)	0.61	0.64	0.68	0.72	0.76	0.79	0.82	0.84	0.86	
Foreign assets ($ millions)										16,500
Domestic Net Investment ($ millions)[a]	*500*	772	1302	1230	1764	2920	3441	5756	5786	23,471
Education expenditure ($ millions)	*40*	86	145	217	387	2029	1338	3528	5341	13,111
Damage from CO_2 emissions ($ millions)	−520	*−540*	*−560*	−600	−660	−700	−780	*−800*	*−800*	−5960
Energy depletion ($ millions)	−1851	−2038	−1814	−2679	−3560	−8117	−8318	−10,715	−25,374	−64,464
Genuine investment rate (%)										**−4.6**
CO_2 emissions (Mt of CO_2)[b]	26	*27*	*28*	30	33	35	39	*40*	*40*	
Crude oil production (in kbd)	**648**	**632**	**568**	**676**	**755**	**765**	**802**	**845**	**843**	
Gas production (in bcm)	***25***	***29***	***31***	***35***	**39**	**45.8**	**50.7**	**63.2**	**76.9**	

Sources: IMF, Ministry of Finance, OPEC, IEA
In yellow, forecasts, estimates, budgets and author's assumptions
CO_2 cost assumed at $20 per ton
[a]Does not take into account financial investments
[b]IEA, and author's own estimates after 1999

Table 10.5 Genuine wealth creation in Qatar (bis)

Qatar	2000	2001	2002	2003	2004	2005	2006	2007	2008	2000-2008
GDP ($ millions)	*16,000*	*19,000*	*20,000*	*25,000*	*35,000*	43,000	57,000	*71,000*	91,000	377,000
Population (millions)	0.61	0.64	0.68	0.72	0.76	0.79	0.82	0.84	0.86	
Foreign assets ($ millions)										16,500
Domestic Net Investment ($ millions)[a]	*500*	772	1302	1230	1764	2920	3441	5756	5786	23,471
Education expenditure ($ millions)	*40*	86	145	217	387	2029	1338	3528	5341	13,111
Damage from CO_2 emissions ($ millions	−520	*−540*	*−560*	−600	−660	−700	−780	*−800*	*−800*	−5960
Energy depletion ($ millions)	−878	−3198	−3418	−3861	−4304	−5057	−5602	−6980	−8497	−41,795
Genuine investment rate (%)										**1.4**
CO_2 emissions (Mt of CO_2)[b]	26	*27*	*28*	30	33	35	39	*40*	*40*	
Crude oil production (in kbd)	**648**	**632**	**568**	**676**	**755**	**765**	**802**	**845**	**843**	
Gas production (in bcm)	***25***	***29***	***31***	***35***	**39**	**45.8**	**50.7**	**63.2**	**76.9**	

Sources: IMF, Ministry of Finance, OPEC, IEA
In yellow, forecasts, estimates, budgets and author's assumptions
CO_2 cost assumed at $20 per ton
[a]Does not take into account financial investments
[b]IEA, and author's own estimates after 1999

Suggestions for future research include ways to genuinely account for the financial investments, particularly given their volatile and exogenous nature. These are unchartered territories as Sovereign Wealth Funds are positioning themselves to separate the "saving funds" function, with a usually large endowment, from more aggressive investments, with typically smaller assets. It is the latter which is typically in charge of investing in riskier ventures or within the country. Also, a better understanding of investments in human capital would be helpful. Finally, it could be misleading to judge the performance of a country over a decade. Ideally, this calculation would be performed over three decades in order to assess the impact on at least one generation. This method could be also used to forecast future genuine wealth creation and support decision making for allocation of resources and capital.

The second major conclusion relates to the pricing of natural resources and fuel subsidies. Assuming these calculations were not imperfect, it would imply that Saudi Arabia for instance would be following a good approach in terms of resources' valuation, by exporting its light crude, by burning its heavy crude and discounted products and by restricting the sales of natural to its domestic market and carefully allocating tranches of supply. It does not mean this path is optimal, but it appears more sustainable than economic policies followed in past periods of high oil prices. Fuel subsidies are meant to support a certain level of economic growth, while exhausting resources at discounted prices without a substantial creation of other forms of capital would not be sustainable.

10.4 Summary

The other major question is whether electricity-sector reforms are robust in view of resource monetization and sustainable development needs. Most MENA countries failed to make genuine investments in the 1970s and 1980s during periods of high oil revenue. Today, it is increasingly admitted that natural resources depletion should be offset by manufactured and human capital increase. But there are still wide discrepancies between countries, particularly between North African and Gulf hydrocarbons exporters. The latter would have benefitted from a "mutualisation (pooling) of interests" while Algeria and Libya endured the existence of vested interests which had time to organize and to benefit from subsidies with no pressure to seek efficiency or genuine investment. It is possible to assess whether countries in MENA are building wealth or becoming impoverished over time, whether their hydrocarbons are properly monetized or whether the depletion of their natural resources is not offset by genuine economic investments. Pricing resources below their "social cost" often lead to inefficient consumption. Others would argue that this under-pricing contributed to industrial and private sector development. The application of the theoretical framework to two major hydrocarbon exporting countries of the region, Saudi Arabia and Qatar, turns that Saudi Arabia, a middle income country according to the World Bank classification, appears to be building wealth more rapidly than Qatar, a high income country. For all the uncertainties and the caveats in the calculation, the two countries seem to be improving the way they manage their economies, and the way they price and manage their resources. The pricing focus could initially be designed to support efficiency-related decisions—combined cycle, grid reinforcement and integration, efficiency programs in industries. Using the fuel supply cost as a minimum reference will be helpful in this regard, without being too punitive for the economy.

Chapter 11
Conclusion: Evolution of the Single Buyer Model in MENA

Abstract The ideal model does not exist, and the State identity evolves. The SBM was so far adequate and the SBM Plus was built here so it fulfils most of the four qualitative criteria used in this evaluation. It is attractive, feasible, credible and remediable. This is not just another study advocating hybrid models as a convenient solution for countries searching for excuses because further deregulation is challenging technically and politically. Trade-offs have to be made between pragmatism and theory. Limited supply leaves the room open for market manipulation, if that market is weakly regulated. Its capital intensity calls for the creation of mechanisms to increase investors' confidence (sovereign guarantees, long-term commitments, take-or-pay clauses) with the State as the backbone of these mechanisms. But governments would opt for the least complex capacity mechanism: direct intervention. Public authorities would then remain involved in the electricity sector, particularly in immature, rapidly growing markets.

Keywords Trade-offs · Hybrid model · Government intervention · Single buyer model

Out of the 18 countries in MENA, 12 are hydrocarbons exporters (8 of which are OPEC countries). For these countries, the robustness of the electric reforms was tested in light of two sets of policy orientations which define their resources monetization directions: electricity fuel mix options and sustainable development needs.

Hydrocarbons dominate the fuel mix, and natural gas, barring a few exceptions, has long been considered as the fuel of choice for power generation. The rich gas reserves position of the MENA region should support that choice. However, it is the deliverability of those reserves which questions the ability of indigenous natural gas to cover all electricity generation needs. The analysis concludes that with current rates of production and development of reserves, natural gas will need to be complemented by other fuel sources. In some countries, oil-fired generation continues to prevail because of unavailability of the gas or because of strategic reasons, but other alternatives are considered.

L. Benali, *Electricity-sector Reforms in the MENA Region*, Perspectives on Development in the Middle East and North Africa (MENA) Region,
https://doi.org/10.1007/978-3-319-96268-9_11

The main alternatives to hydrocarbons in electricity generation in the region are coal, renewables and nuclear. Another alternative source of supply is the "hidden" generation or demand-management. All these complementing alternatives pose institutional, financial and practical issues.

Assuming the same gas price maintained at around \$1–1.5 per MMbtu, other alternatives can not compete on a pure cost basis. Moreover, with the exception of coal, other alternatives will have to be directly or indirectly supported by the State, but these could act as a hedge against uncertain hydrocarbons supplies or missed CO_2 "opportunity charges". Nuclear carries strategic weight but its development in the region would require heavy institutional, regulatory, educational and diplomatic investments and will take a long time to materialize. Even in situations where hydrocarbons reserves and supplies are abundant, at a stable cost, subsidies growth, because of growing demand, would lead to a reconsideration of fuel pricing policies. But the reality is that, ultimately, the price will have to evolve, as marginal fuel production costs are increasingly higher than fuel prices agreed under several Fuel Supply Agreements in the region. Evolving fuel prices could be viewed, at first sight, as challenging for any introduction of an additional degree of deregulation, but there are various steps, including a period of assessment and testing, that should be followed before price changes. New fuel pricing can be applied in varying degrees, depending on the emphasis and the impact sought. It could be applied to support efficiency-related initiatives, to discriminate between end-use sectors or to encourage fuel displacement. The transition is meant to accompany the country as it transforms from a fuel exporter to a fuel monetizer and then to a mature economy.

Electricity-sector reforms are not in contradiction with a fuel diversification policy, but introducing new fuels and new technologies will require the State to step in directly or indirectly—through the upstream sector or the investment arms. This is also relevant for green initiatives and renewables, as having a centralized policy planning and a high institutional feasibility should be an enabler. Regulation will have to play a critical role to ensure that the electricity cost structures and the efficiency of the system are maintained or enhanced, and that the right investment signals are sent at the right time. Changes in fuel offerings (fuel type, pricing, contract terms...) will require a delicate balancing act. They will test all already established market models in the region and will challenge decisions to deregulate further.

The other major question to answer is whether electric reforms are robust in view of resource monetization and sustainable development needs. It conventionally admitted that most MENA countries failed to make genuine investments in the 1970s and 1980s during periods of high oil revenue. Today, this is changing with the increased realization that natural resources depletion should be offset by manufactured and human capital increase. But there are still wide discrepancies between countries, particularly between North African and Gulf hydrocarbons exporters. The latter would have benefitted from a "mutualization of interests" while Algeria and Libya endured, among other things, the existence of vested interests which had time to organize and to benefit from subsidies with no pressure to seek efficiency or genuine investment.

In any case, it is possible to assess whether countries in MENA are building wealth or becoming impoverished over time, whether their hydrocarbons are properly monetized or whether the depletion of their natural resources is not offset by genuine economic investments. Pricing resources below their "social cost" often lead to inefficient consumption. Others would argue that this underpricing contributed to industrial and private sector development. In today's conditions, long term fuel supply agreements with below-market prices are questioned sometimes renegotiated. Contributions of the various "benefiters" to the economy are scrutinized and evaluated.

The application of the theoretical framework to two major hydrocarbon exporting countries of the region, Saudi Arabia and Qatar, turns surprising results. Saudi Arabia, a middle income country according to the World Bank classification, appears to be building wealth more rapidly than Qatar, a high income country. Understanding all the caveats inherent to the calculation is important. For example, Qatar being more vulnerable to the pricing of gas than Saudi Arabia is to oil, there is a risk of understating or overstating wealth creation in Qatar. For all the uncertainties in the calculation, the two countries seem to be improving the way they manage their economies, and the way they price and manage their resources. The last statement is particularly true for Saudi Arabia which exports its light crude, consumes its heavy crude and discounted products and sells its gas domestically at below-market price.

Once these models enhanced and reinforced with additional data and inclusion of other components, it would be interesting to run sensitivity analysis based on fuel pricing and monetization. The impact of various fuel prices on GDP will be central, particularly on industries that constitute a major source for fuel demand (aluminium, cement, petrochemicals, electricity ... etc).

The added value of exports depends on export prices and actual amounts of exports. Decreasing consumption would free up more fuel for exports. Gradual change in price could first drive efficiency in energy usage, encourage industrial and LNG users to seek alternative fuel or feedstock, then force consistent and efficient choices in new energy-related investments and to respond to real marginal market values.

The pricing focus could initially be designed to support efficiency-related decisions—combined cycle, grid reinforcement and integration, efficiency programs in industries. Using the fuel supply cost as a reference will be helpful in this regard, without being too punitive for the economy. Indeed, below a certain ceiling, fuel price increase (to electricity generators and industrials) will have a gradual impact on energy intensity (I) without impacting GDP (U). If more price impact or fuel displacement is needed, pricing could rise to support the next expensive option (fuel alternative or market price). In this case, GDP will be directly affected. For example, export-oriented industries might relocate production to other countries. Conversely, that would free up more hydrocarbons for exports, potentially increasing the government's direct revenues. Some would argue that this development would mean a return to the old era of *rentier states*. However, in hydrocarbons producing countries where fuel supplies are becoming tight, a redistribution of GDP might be necessary. Instead of energy-intensive industries, economy might be

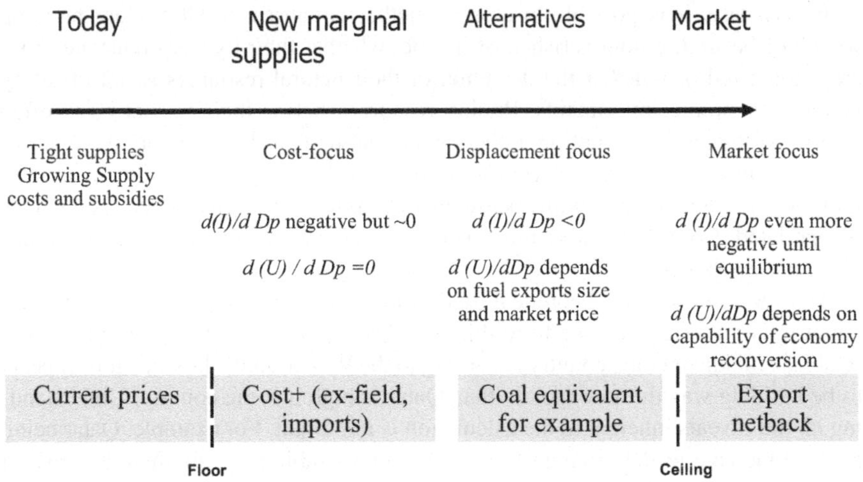

Fig. 11.1 Energy intensity and economic growth reaction to changing fuel prices

reoriented towards less energy-intensive services. For example, Dubai or Bahrain, which were suffering from dwindling oil output, tried to turn their countries into financial hubs for the region.[1] Today, a significant part of these economies is based on the financial-related sector.[2] The downside is that this makes them more vulnerable to a global economic slowdown than their neighbours and to market prices when they tried to reform domestic prices (Fig. 11.1).

The macroeconomic reorientation could however be long and costly, even painful. Decreasing drastically the energy intensity of an economy opens economic, political and more importantly, social issues.

Research has been developed on the link between energy prices and energy intensity. Most analyses have been done ex-ante and are based on econometric models. Shi and Polenske (2006) for example used various economic models (input-output-based structural decomposition analysis (SDA) and econometric regression models and by using a decomposition method of own-price elasticity of energy intensity) in their analysis of China's energy intensity. Among other things, they observed a greater sensitivity (in terms of the reaction of energy intensity towards changes in energy prices) of the industry sector, compared to the overall economy.

Simulations on the impact of price changes (at least to match future supply costs levels) on energy intensity, on GDP and then on wealth creation and genuine investment rates could guide future policy decisions regarding investments, macroeconomic orientations and fuel pricing. Differences could be made on the fuel price

[1]Some would use the Singapore model of development as a reference for Dubai and Bahrain's development.

[2]Fourteen percent in Dubai, 25% in Bahrain.

to be paid by the industries and by the electricity sector, depending on the goal pursued.

> Someone who is regulated is never going to be a "happy camper". There are always going to be feelings of abuse, of unnecessary interference, of a lack of productivity, and sometimes, of outright incompetence (Michael Greenberger, Professor at the University of Maryland School of Law, former official at the CFTC)

Reform is considered as an evolutionary process that accompanies cultural and mentality change, it does not precede it. At times, it has to. Electric reforms might be viewed, particularly by opponents, as the "grand victory of capitalism" or "the triumph of markets over governments". However, they merely boil down to an ordinary financial question, where private interests overlap with national needs. The reality is that regulation is still needed.

Electricity is a magnet for political intervention. It is perceived as an essential public good, ubiquitous and unavoidable. All consumers are direct or indirect stakeholders, in the sense that, even in developing countries, access to the service is a priority, sometimes at low or no fee. To add to the complexity, electricity can not be economically stored or transportable on long distances with the current and foreseeable state of technology, contrary to oil or gas. Therefore, it might be one of the rare "commodities" to justify the creation of regional markets. Moreover, as with commodities, limited supply leaves the door open for market manipulation, if that market is weakly regulated. Electricity's capital intensity calls for the creation of mechanisms to increase investors' confidence (sovereign guarantees, long-term commitments, even take-or-pay clauses). It is usually the State which serves as the backbone of these mechanisms. Last but not least, there could be significant demand and supply adjustment lags if price was left to do all the economic work. Because governments would find supply interruptions (and price volatility) unacceptable, they would opt for the least complex capacity mechanism of all: direct intervention.

Public authorities would then have every good reason (technical and political) to remain involved in the electricity sector, particularly in immature, rapidly growing markets like in MENA, with tight reserve margins. In Europe, with the notable exceptions of Spain and Italy, liberalization was initiated during periods of significant excess capacity. Low energy prices and low costs of new supplies favoured new entrants, setting their costs below the average cost of generation, the typical metric for setting regulated tariffs. Moreover, the relatively large size supported the development of liquid wholesale markets. Then, the question for MENA countries is: where to set the limit of policymakers' intervention?

Let's be fair and realistic. It is better to fulfil a few criteria than none. National champions in MENA, which would be dual firms, ideally having the political weight and the economic pragmatism to survive, are not necessarily a problem, providing strong regulation exists. Remediability might be jeopardized but consumer protection and the political structure of MENA countries should balance that. Are national champions unavoidable in MENA? So they are probably, given the multiple identities and roles of the State. Here again, academics are increasingly "accepting" market power as long as the market is contestable.

It was demonstrated that high oil prices do not necessarily freeze or reverse reforms in oil-producing countries. Actually, high oil revenues were invested locally and fuelled demand growth. This created the need for robust models to induce investments while keeping governments involved. *Long-term coordination and long-term commitment seem crucial. High oil revenue post-2001 also created new financing and structure possibilities, even in countries hitherto closed to IPPs, such as Libya or Kuwait*. Until recently, the Libyan electricity sector was still organized around a stated-owned vertically integrated monopoly (GECOL), and was still struggling with the same budgetary issues that have long plagued the development of the sector. The country was nonetheless seeking to open up for IPPs, but the attractiveness, the feasibility and the credibility of such schemes were questioned. Then, GECOL, eager to make projects into IPPs decided to sell one plant (Tripoli West) once it is built, to an already defined buyer (UAE-based Oasis IP) that will sell the electricity generated to GECOL. The plan was to involve its own SWF, the Libyan Investment Authority, and another hydrocarbons' exporter, the government of Abu Dhabi, to finance the deal. The regime change and the consequent war in the country froze reform efforts, but the momentum was there. In Kuwait, the Partnerships Technical Bureau, created under legislation in 2008, oversaw the country's first PPP, an IWPP. The Electricity and Water Ministry acts as the buyer of the output.

On the other hand, Saudi Arabia and others who have already introduced competition are reviewing their reform programs, in light of new international, regional and domestic developments. *Evolution is necessary. The issue is nonetheless still the same: how to weigh short term tangible costs against long-term hypothetical benefits?* Time is a critical factor and rigid structures tend to produce long reforms, but which would be, in turn, difficult to reverse.

There are hardly any countries in the region that could have the luxury of delaying investment decisions in generation. Those who can would invest their resources to implement efficiency programs, demand-side techniques and could support alternatives that would not compete with thermal generation otherwise (renewables, nuclear). The majority is facing difficult trade-offs: trade-offs in investments and budgets, in allocation of resources (human, institutional and natural), in the tasks devoted to the State and in the identities of the State.

Generally in MENA, the identity of the State evolves: from a Rentier State to a Developer State, and from a fuel exporter to a fuel monetizer. The involvement of public authorities in electric reforms evolves with it. However, this process is too slow in comparison with the major challenges facing the region. *Fuel supplies' tightness and development needs are gradually pressuring stakeholders and are necessarily being taken into account*. Fuel pricing structures are being reviewed, energy fuel mixes are being analyzed, industrialization policies are being scrutinized, etc. The results are mixed but there are some efforts, albeit slow, in the right direction.

There are nonetheless two more looming challenges that are often overlooked and that lead us to question the sustainability of the electric market designs adopted in the region. First, building generation capacity is important to address the high demand growth experienced by the region, but market development would require additional investments in networks' adequacy and robustness. The latter are necessary conditions to accompany the growth of the market and to enable the additional contractual/regulatory arrangements defined as components of the SBM Plus. Second, future contractual/regulatory arrangements and structure will need to take into account the potential for demand growth deceleration. For example, bilateral contracts would decrease the offtake risk, reducing the need for the government to bear most risks. In the same vein, contracts could be shorter in time (and therefore with less liability) to reflect the changing pattern of demand growth.

In this regard, there is an established consensus that the "ideal" model, to be duplicated in several countries, does not exist. Also, there is an increasing academic consensus that a fair amount of flexibility in the models was needed. The evaluation concluded that the SBM was *so far* suitable in most countries in MENA. Its implementation is probably not optimal given the limited impact on costs and investments noticed. Given all the current and future challenges mentioned above, some form of evolution is needed. The SBM Plus was built so that it fulfils most of the four qualitative criteria used in this evaluation. It is attractive since it builds on the advantages of the SBM. It is feasible because it does not call for major additional institutional or industrial changes. It is credible in the sense that it reinforces a centralization of planning and a decentralization of risks. It is also credible since it would facilitate fuel and energy pricing revisions (particularly if contracts are shorter). Finally, policy-makers should ensure that enough flexibility is embedded in the model to ensure its remediability, despite the willingness to create, recreate or strengthen a national champion. The importance of independent regulation, alongside predictability of decisions, as prerequisites for the reform's credibility, was already emphasized. The national champion falling outside of government control is an additional reason to guarantee the regulator autonomy from political intervention and high profile, and to limit information asymmetry to the extent possible.

This is not just another study advocating hybrid models as a convenient solution for countries searching for excuses because further deregulation is challenging technically and politically. Trade-offs have to be made between pragmatism (evolution) and theory (status quo). This analysis takes into account practical limitations of textbook reforms and deregulation. These restrictions are embedded in the characteristics of these markets (size, availability of human resources, costs, political capital). They also derive from lessons learnt from deregulation experiences in other parts of the world. The institutional approach has been helpful in order to evaluate the various current and potential models in the region, but additional aspects such as the resource endowment or the exogenous factors had to be overlaid.

At the same time, one major challenge facing the energy sector in general and the electricity sector in particular is to find a path of economic, environmental and social sustainability. The combination of a neo-institutional approach with a methodology inspired from welfare economics has been helpful in that regard. Market forces have demonstrated that they could be adequate levers to stimulate innovation and achieve a more efficient allocation of resources. However, sustainability can not be left to them. More particularly, environmental and social considerations would also need complementary instruments, fulfilled by governments, in the absence of a better solution.

Another underlying goal in this study was to revisit the determinism of the Dutch Disease on MENA oil exporting countries and the general assumption of low economic performance automatically associated with oil producing countries. Indeed, Contrary to historical trends, some MENA hydrocarbons producers appear to be building genuine wealth, but probably not the way one would think. The aggressive fuel pricing policies have been helpful in building an industrial base, and in stimulating the non-oil economy. However, there is no indication whether this path is optimal. Moreover, fuel pricing is evolving to a new form anyway, given the existing and looming challenges. Countries are facing trade-offs in the way they plan their electricity investment programs, as explained earlier. They also face trade-offs in their economic orientations. Their approach to price their indigenous resources domestically has a direct impact on these orientations. It is important to keep in mind that private power projects, or associated industrial projects, are supposed to redistribute the government's rents differently, but generally without affecting end-user tariffs. For example, when the company is listed in the stock exchange and performing with the benefit of low fuel prices, local investors are expected to benefit. Moreover, privatization and IPPs can also be meant to create prospects for local entrepreneurs, in order to develop personal wealth. In all cases, even if fuel and electricity tariffs are subsidized at various levels, and end-users are insulated from costs' variations, consumers are actually stakeholders in the process. The triangle of opportunism (government, companies and consumers) remains relevant in MENA despite continuing subsidies.

If subsidies' lifting were to be considered, this change would have to be part of a more comprehensive economic reform. The latter would affect various parts of the economy, in addition to the energy sector. Obviously, it would be politically sensitive, especially since economic reforms are generally substitutes for political reforms in the region (Chinese model). But before lifting subsidies, new fuel pricing can be applied in varying degrees, depending on the emphasis and the impact sought. It could be applied to support efficiency-related initiatives, to discriminate between end-use sectors or to encourage fuel displacement. The transition is meant to accompany the country as it transforms from a fuel exporter to a fuel monetizer and then to a mature economy. Here again, electric reforms are not in contradiction with a fuel diversification policy, but introducing new fuels and new technologies could require the State to step in again directly or indirectly—through the upstream sector or the investment arms.

In addition to new fuels and new generation technologies, it is clear that the region is in need of efficiency-improvement techniques and demand-management programs. The genuine wealth creation would be accelerated. If the introduction of these programs were to be left to market forces only, the price of the marginal kWh would need to be substantially higher than average wholesale prices, in order to encourage new entrants. Therefore, introducing an additional level of sophistication in the form of clean technologies, new types of transportation, or even smart grids, still requires a substantial level of government support, at least in the design and launch phase.

11.1 Summary

This chapter reemphasizes that the ideal model, duplicable to several countries, does not exist. At the same time, the State identity would evolve: from a Rentier State to a Developer State, and from a fuel exporter to a fuel monetizer. The SBM was so far adequate in most MENA countries. The SBM Plus was built so that it fulfils most of the four qualitative criteria used in this evaluation. It is attractive since it builds on the advantages of the SBM. It is feasible because it does not call for major additional institutional or industrial changes. It is credible in the sense that it reinforces a centralization of planning and a decentralization of risks. It is also credible since it would facilitate fuel and energy pricing revisions (particularly if contracts are shorter). Finally, policy-makers should ensure that enough flexibility is embedded in the model to ensure its remediability, despite the strong temptation to create, recreate or strengthen a national champion, and for the government to continue to be involved in the sector and its pricing administration.

This is not just another study advocating hybrid models as a convenient solution for countries searching for excuses because further deregulation is challenging technically and politically. Trade-offs have to be made between pragmatism and theory. This analysis takes into account practical limits of reform models and deregulation, bearing in mind that electricity is a magnet for political intervention. The sector is perceived as an essential public good, ubiquitous and unavoidable. Limited supply leaves the room open for market manipulation, if that market is weakly regulated. Its capital intensity calls for the creation of mechanisms to increase investors' confidence (sovereign guarantees, long-term commitments, take-or-pay clauses) with the State as the backbone of these mechanisms. Last but not least, there could be significant demand and supply adjustment lags if price was left to do all the economic work and because supply interruptions (and price volatility) would be unacceptable to governments, the latter would opt for the least complex capacity mechanism of all: direct intervention. Public authorities would then have all the good reasons to remain involved in the electricity sector, particularly in immature, rapidly growing markets.

Acronyms and Units

ADWEC/ADWEA	Abu Dhabi Water Electricity Company/Authority
BCM	Billion Cubic Meters
BOT/BOOT/BOO	Build Own Transfer/Build Own Operate Transfer/Build Own Operate
BST	Bulk Supply Tariff
BTU	British Thermal Units
CAGR	Compound Average Growth Rate
CCGT/GT/CC	Combined Cycle Gas Turbine/Gas Turbine/Combined Cycle
DSM	Demand Side Management
DuoS	Distribution Use of System
EIA	Energy Information Administration
EPC	Engineering, Procurement, Construction
FSA	Fuel Supply Agreement
HFO	Heavy Fuel Oil
IEA	International Energy Agency
IOC/NOC	International/National Oil Company
IPP/IWPP	Independent Power Project/Independent Power and Water Project
LNG	Liquefied Natural Gas
MEED	Middle East Economic Digest
MEES	Middle East Economic Survey
MMgd	Million Gallons per Day (Water)
O&M	Operations and Maintenance
PPA/PWPA	Power (and Water) Purchasing Agreement
SMP	System Marginal Price
Solar PV	Solar Photo Voltaic
SRMC/LMRC	Short (Long) Run Marginal Cost
TuoS	Transmission Use of System

L. Benali, *Electricity-sector Reforms in the MENA Region*, Perspectives on Development in the Middle East and North Africa (MENA) Region,
https://doi.org/10.1007/978-3-319-96268-9

Appendix A: Temperatures for Air Conditioning, Middle Eastern Versus US Cities

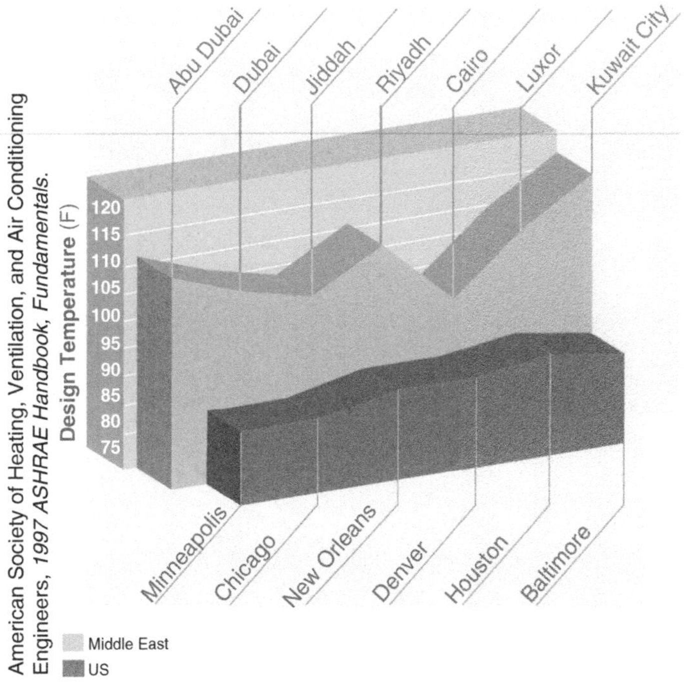

L. Benali, *Electricity-sector Reforms in the MENA Region*, Perspectives on Development in the Middle East and North Africa (MENA) Region,
https://doi.org/10.1007/978-3-319-96268-9

Appendix B: The Four Organizational Models of the Electricity Sector

L. Benali, *Electricity-sector Reforms in the MENA Region*, Perspectives on Development in the Middle East and North Africa (MENA) Region,
https://doi.org/10.1007/978-3-319-96268-9

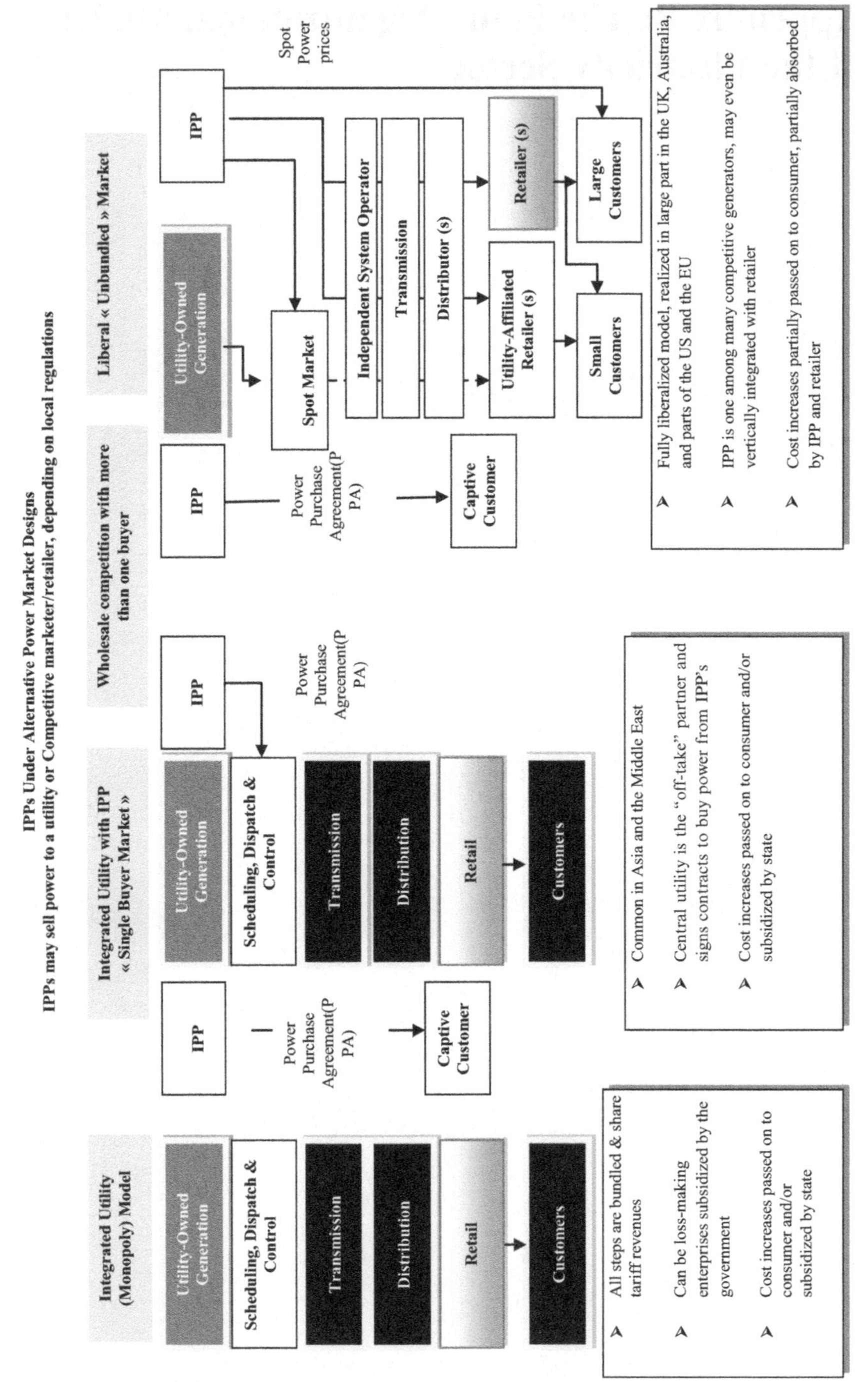

IPPs Under Alternative Power Market Designs
IPPs may sell power to a utility or Competitive marketer/retailer, depending on local regulations
Integrated Utility (Monopoly) Model
Utility-Owned Generation
Scheduling, Dispatch & Control
Transmission
Distribution
Retail
Customers
All steps are bundled & share tariff revenues
Can be loss-making enterprises subsidized by the government
Cost increases passed on to consumer and/or subsidized by state
Integrated Utility with IPP « Single Buyer Market »
IPP
Power Purchase Agreement(PPA)
Captive Customer
Utility-Owned Generation
Scheduling, Dispatch & Control
Transmission
Distribution
Retail
Customers
IPP
Power Purchase Agreement(PPA)
Common in Asia and the Middle East
Central utility is the "off-take" partner and signs contracts to buy power from IPP's
Cost increases passed on to consumer and/or subsidized by state
Wholesale competition with more than one buyer
IPP
Power Purchase Agreement(PPA)
Captive Customer
Liberal « Unbundled » Market
Utility-Owned Generation
IPP
Spot Market
Independent System Operator
Transmission
Distributor (s)
Utility-Affiliated Retailer (s)
Retailer (s)
Small Customers
Large Customers
Spot Power prices
Fully liberalized model, realized in large part in the UK, Australia, and parts of the US and the EU
IPP is one among many competitive generators, may even be vertically integrated with retailer
Cost increases partially passed on to consumer, partially absorbed by IPP and retailer

Appendix C: Euro-MENA Grid and Renewables

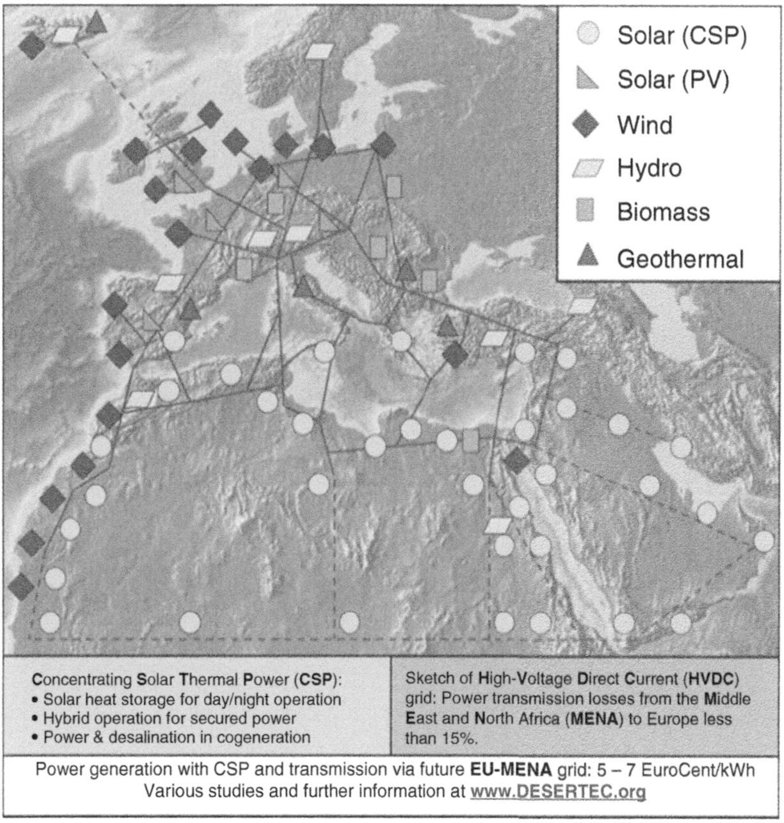

L. Benali, *Electricity-sector Reforms in the MENA Region*, Perspectives on Development in the Middle East and North Africa (MENA) Region,
https://doi.org/10.1007/978-3-319-96268-9

Appendix D: MENA Military Defense Budgets

Country	Defense budgets 2005	2006	2007	2008
Algeria	2870	3090	4270	–
Bahrain	463*	478*	539*	552
Egypt	260	3040	3160	–
Iran	5200	6480	7451	–
Iraq	1650**	–	–	–
Israel	7670	7530	9260	–
Jordan	973	1100	1620	1940
Kuwait	4400	3490	3750	4930
Lebanon	638	588	733	751
Libya	721	593	656	–
Mauritania	18	18	19	–
Morocco	2070	2160	2400	2770
Oman	3690	3270	3230	–
PA	–	–	–	–
Qatar	2190est	2330est	109	–
S. Arabia	25,400	29,540	35,400	38,200
Syria	1250	1730	1460	–
Tunisia	430	443	470	–
UAE	2560	9480	10,080	–
Yemen	583	823	908	–
Total* **	**63,726**	**76,183**	**86,496**	–

Source: IISS, The Military Balance 2009, The Military Balance 2008, and The Military Balance 2007, MEES
MENA Military Defense Budgets, 2005–2008 ($Mn)
* Excluding extra budgetary funding
** Figures not from IISS; estimates made in Iraq following publication of the Coalition Provisional Authority's 2004–2006
Budget (MEES, 12 April 2004)
*** Excluding Iraq

L. Benali, *Electricity-sector Reforms in the MENA Region*, Perspectives on Development in the Middle East and North Africa (MENA) Region,
https://doi.org/10.1007/978-3-319-96268-9

Appendix E: The Example of Jordan's Renewable Policy

The government's target is 5% of all energy produced to be from renewable energy by 2015.

Draft Renewable Energy Law No. 2007 (based on a study prepared by German consultancy company Lahmeyer International. Initial study funded by a Japanese grant). The Draft reviewed by Electricity Regulatory Commission, National Electric Company and Jordanian National Centre for Energy Studies. Draft processed by the Legislation and Opinion Bureau before being submitted to Parliament. Expected to be passed as law quickly as the Parliament is under pressure to take immediate action to combat rise in price of oil.

Key Provisions

- Defines *renewable energy sources: the energy produced by natural and non-traditional sources including the solar energy, wind energy, underground energy, hydrogenic energy, bio-energy (bio fuel) and energy saving.*
- Ministry to locate and protect for development suitable land areas for renewable energy technology (Art 5).
- Ministry to receive and approve offers to develop renewable energy plants on the land and issue a Renewable Energy Source Certificate to Qualified Renewable Energy Plants (Art 11).
- Licensees authorised to carry out wholesale or retail supply or distribution, as well as main consumers, shall accept uploading of the electricity produced from a Qualified Renewable Energy Plant, as long as it is offered as per the Law's tariff provisions in Art 14, notwithstanding the priority order specified in the Network Code or any other uploading rules (Art 13).
- Incentives: 75% income tax exemption for the first 10 financial years on the profits of Qualified Renewable Energy Plants (Art 15); design, engineering, construction, financing, operation, maintenance, licensing agreements and contracts, as well as sale agreements exempted from revenue stamps and university fees (Art 16).

L. Benali, *Electricity-sector Reforms in the MENA Region*, Perspectives on Development in the Middle East and North Africa (MENA) Region,
https://doi.org/10.1007/978-3-319-96268-9

- Expropriation of land permitted and leased at an "encouraging fee" (Art 17).
- Government investment and proceeds from Clean Development Mechanism of Kyoto Protocol allocated to reduce tariff on electrical power produced by Qualified Renewable Energy Plants (Arts 19 and 20).
- Fund for Renewable Energy and Rationalisation of Energy established to take part in developing renewable energy sources (Art 21).

Potential Shortcomings

- Minimum 5 year experience threshold.
- Insufficient incentives.

Appendix F: Transmission Map of MENA

L. Benali, *Electricity-sector Reforms in the MENA Region*, Perspectives on Development in the Middle East and North Africa (MENA) Region,
https://doi.org/10.1007/978-3-319-96268-9

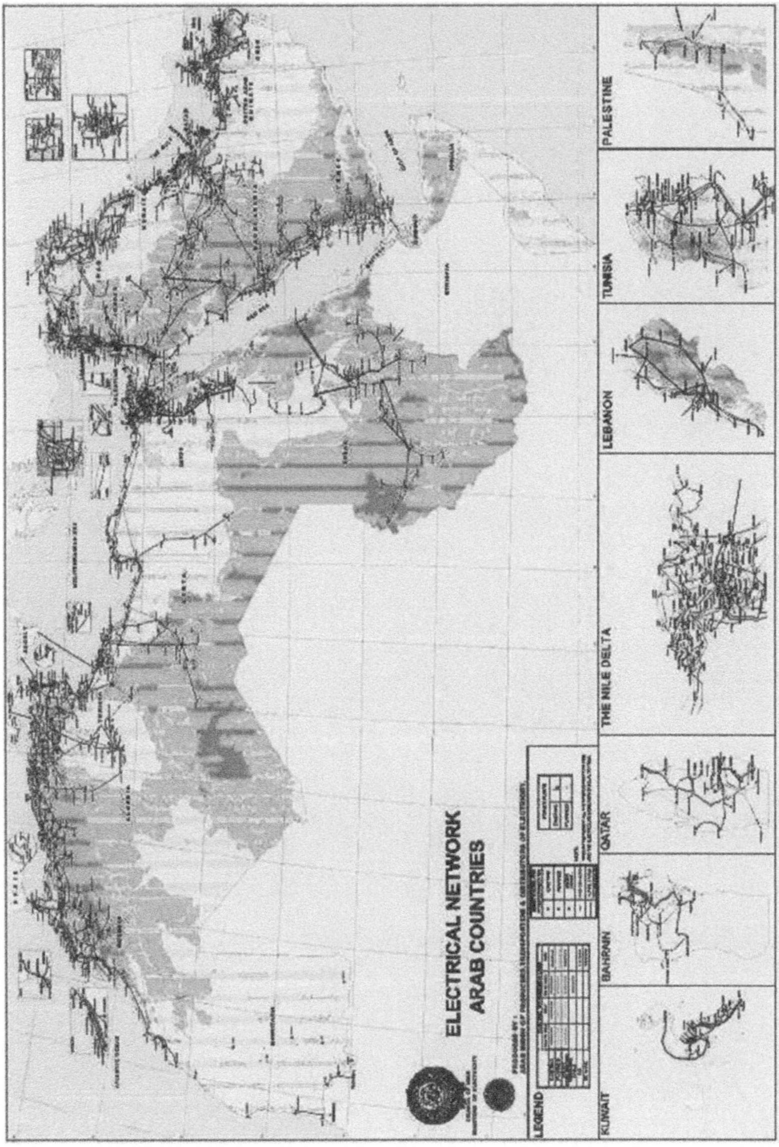
ELECTRICAL NETWORK
ARAB COUNTRIES
LEGEND
KUWAIT
BAHRAIN
QATAR
THE NILE DELTA
LEBANON
TUNISIA
PALESTINE

Appendix G: Middle Eastern Sovereign Wealth Funds

Country	Fund	Assets (US $bn)	Established	SWFI transparency rating (out of 10)
UAE	Abu Dhabi Investment Authority	875	1976	3
Saudi Arabia	SAMA Foreign Holdings	~365	n/a	2
Kuwait	Kuwait Investment Authority	~265	1953	6
Qatar	Qatar Investment Authority	60	2003	5
UAE	Investment Corporation of Dubai	>18	2006	4
Iran	Oil Stabilisation Fund	13	1999	1
Bahrain	Mumtalakat Holding Company	10	2006	6
UAE	Mubadala Development Company	10	2002	6
Oman	Oman State General Reserve Fund	10	1980	1

Source: SWFI (2008)
This is a non-exhaustive list

L. Benali, *Electricity-sector Reforms in the MENA Region*, Perspectives on Development in the Middle East and North Africa (MENA) Region,
https://doi.org/10.1007/978-3-319-96268-9

References

Aissaoui A (2008) Inflation in MENA region: causes, policy responses and challenges. APICORP Res Econ Comment 3(9), September

Al Khatib H (2000) Regional electricity forecasts and challenges to 2020. Mees XLIII(10), March 6

Al Maktoum M (2009) Education vs. extremism. The Wall Street Journal, New York, June 3

Al Naimi A (2009) Achieving energy stability in uncertain times. CERAWeek Keynote Address, Houston, February 10

Alami R (2005) Financial aspects of Arab power development. Oxford Institute for Energy Studies, Oxford

Almeida ELF, de Oliveira A (2000a) Developing the gas industry in Brazil: is it possible under competitive regulatory framework? 23rd annual international conference of the IAEE, Sydney, June 7–10

Almeida ELF, de Oliveira A (2000b) Regulatory issues in Brazilian emerging gas industry. Rio Oil and Gas Expo Conference, October

Amr AI (2006) Temporary substitution to independent regulation in the electricity sector—application of international practices. Paper Presented at Power-Gen Middle East

Anthioumane T, Ndiaye M (2005) Vent de Panique sur le Trésor? La Gazette du Maroc, January

Arab Union of Producers Transporters & Distributers of Electricity (2008) At'arifate al ka'hrabaiya fi al watane al'arabi. [Les tarifs de l'électricité dans les pays arabes]. Commission de planification, November

Arentsen M, Dunneke R (1996) Economic organisation and liberalisation of the electricity industry. Energy Policy 24(6)

Arrow K, Dasputga P, Goulder L (2003) Are we consuming too much? sl

Arrow K et al (2004) Are we consuming too much? J Econ Perspect, Am Econ Assoc 18(3)

Arrow K, Dasputga P, Goulder L (2008) The US and China sustainability: perspectives based on comprehensive wealth. sl

Artus P, Betbeze JP, de Boissieu C et al (2003) « 5+5 »: l'ambition d'une association renforcée

Atallah TN, Hunt LC (2016) Modeling residential electricity demand in the GCC countries. KAPSARC, April

Averch H, et Johnson LL (1962) Behavior of the firm under regulatory constraint. Am Econ Rev 52 (5):1053–1069, December

Bacon R (1995) Privatization and reform in the global electricity supply industry. Annu Rev Energy Environ 20:119–143

Bacon R (1999) Global energy sector reform in developing countries: a scorecard. Report no 219–299, UNDP/World Bank, Washington

L. Benali, *Electricity-sector Reforms in the MENA Region*, Perspectives on Development in the Middle East and North Africa (MENA) Region,
https://doi.org/10.1007/978-3-319-96268-9

Bacon R, Besant-Jones J (2000) Global electric power reform, privatization and liberalization of the electric power industry in developing countries. Annu Rev Energy Environ 26:331–359
Bacon R, Besant-Jones J (2002) Global electric power reform, privatization and liberalization of the electric power industry in developing countries. The World Bank Group Energy & Mining Sector Board, Paper no 2, June
Barrett A, Benali L (2007) Energy pinch in the energy heartland. Wall Street J, New York, 7 Feb
Barrett A, Benali L (2008) Energy pinch in the energy heartland. Wall Street Journal, New York, February
Baumol W (1982) Contestable markets: an uprising in the theory of industrial structure. Am Econ Rev, no 72
Beblawi H, Luciani G (1987) The Rentier state in the Arab world. Croom Helm, London
Beesley ME, Littlechild SC (1989) The regulation of privatized monopolies in the United Kingdom. Rand J Econ 20(3):417–436
Benali L (2002) Concessions et privatisations de l'électricité et de l'eau aux Emirats Arabes Unis: L'évolution encadrée vers l'interventionnisme libéral? Institut d'Etudes Politiques, Mémoire de DEA, Paris
Benali L (2004a) Comment faire partie du jeu? Investir dans l'énergie au Moyen-Orient, Accomex, February 04
Benali L (2004b) Iran: l'équilibre fragile d'un bilan énergétique, in Chevalier. p 337
Benali L (2008) Energy pinch in the energy heartland. Wall Street Journal (see Barrett)
Benhima D (1999) Les Réformes du Secteur Electrique: Le Cas du Maroc. Liaison Energie-Francophonie 44:33–35
Blumsack S, Lave LB, Ilic M (2007) A quantitative analysis of the relationship between congestion and reliability in electric power networks. Cleveland, USA: International Association for Energy Economics (IAEE). Energy J 28(4):28
Boisseleau F (2005) The role of power exchanges for the creation of a single European electricity market: market design and market regulation. Lavoisier, Paris
Borenstein S (2000) Understanding competitive pricing and market power in wholesale electricity markets. Electr J 13
Borenstein S, Bushnell J (1998) An empirical analysis of the potential for market power in California's electricity industry. NBER Working Paper No w6463, March
Borenstein S, Bushnell J (2000) Electricity restructuring: deregulation or reregulation? Regulation, Bushnell 23(2)
Borenstein S, Bushnell J, Stoft S (1997) The competitive effects of transmission capacity in a deregulated electricity industry. NBER Working Paper No w6293, November
Borenstein S, Bushnell J, Wolak F (2000a) Diagnosing market power in California's restructured wholesale electricity market. NBER Working Paper No w7868, September
Borenstein S, Bushnell JB, Wolak FA (2000b) Diagnosing market power in California's deregulated wholesale electricity market. University of California Energy Institute Power Working Paper PWP-064r
Bouayad B (2001) Privatisation du Secteur de l'Electricité au Maroc: Evaluation à l'aide de l'Approche du vote Majoritaire. (a Majority Voting Approach). Marrakech, Maroc: Université Cadi Ayyad, Québec, Canada: Green Université Laval. p 20. Papier présenté au Colloque International du Réseau MONDER « Mondialisation, Energie, Environnement », Paris X, June 13
Bouissou JM (2000) L'économie politique de dix ans de crise. Economie Internationale, no 84, 4ème trimestre
Bowen BH, Sparrow FT, Yu Z (2002) Policy analysis in the development of integrated middle east regional energy markets
Brown D (2000) An overview of electric deregulation lessons learned from overseas. National Utility Service (NUS), news&analysis@poweronline
Brunekreeft G, Keller K (2001) The electricity supply industry in Germany; market power or power of the market? Util Policy 9(1):15–29

Brunekreeft G, Mcdaniel TM (2005) Policy uncertainty and supply adequacy in electric power markets. Oxf Rev Econ Policy 21(1):111–127

Brunekreeft G, Newbery DN (2006) Should merchant transmission investment be subject to a must-offer provision? J Regul Econ 30(3):233–260

Busch A (2009) Same game, different rules. Energy Compass XX(2), January 9

Cadwalader MD, Harley SM, Pope SL et al (1998) Reliability, scheduling markets and electricity pricing. Cambridge, Massachusetts 02138

Cai H, Treisman D (2002) Globalization and corruption: does competition for capital discipline governments? University of California, Department of Economics, Department of Political Science, Los Angeles

Carlsson F, Martinsson P (2007) Willingness to pay among Swedish households to avoid power outages: a random parameter Tobit model approach. Energy J 28(1), IAEE

Carreon V, Jimenez A, Rosellon J (2003) The Mexican electricity sector: economic, legal and political issues. Working Paper No 5, The Program on Energy and Sustainable Development at Stanford University, CA, November

Cartelier L (1998) Existe-t-il un fondement théorique à la notion de service public? Sociétés Contemporaines, no 32, pp 25–35

Center For Administrative Innovation In The Euro-Mediterranean Region (CAIMED) (2006) Regulatory activities in the energy sector of analyzed countries of the Mediterranean, May

CERA (2005) In search of reasonable certainty: oil and gas reserves disclosure. Special report

Champsaur P (2009) Rapport de la Commission sur l'organisation du marché de l'électricité, April

Chevalier JM (2001) Enjeux énergétiques en Méditerranée, Université Paris IX Dauphine, Centre de Géopolitique de l'Energie et des Matières Premières, cahier de recherche 02/2001

Chevalier JM (2002) Energy: the security of supply in question. IOA Newslett 13(3), Autumn

Chevalier JM (2004) Les Grandes Batailles de l'Energie, petit traité d'une économie violente. Gallimard, Paris

Chevalier JM, Percebois J (2007) Gaz et électricité: un défi pour l'Europe et pour la France. Marchés européens de l'électricité et du gaz. Quels prix? Quelle marge de manœuvre pour la France? Rapport, CAE: Conseil d'Analyse Economique

Chevalier JM, Salaun F (1995) Recomposition des industries électriques: internationalisation, nouveaux entrants, diversification. Revue de l'Energie, no 465

Coase RH (1960) Le problème du coût social, Revue française d'économie 7(4)

Commission on the Measurement of Economic Performance and Social Progress (2008). See http://www.stiglitz-sen-fitoussi.fr

Cordesman A (2003) Saudi Arabia enters the twenty first century. Praeger Publishers, Westport

Crampton P, Stoft S (2005) A capacity market that makes sense. Electr J 18:43–54, August/September

Cramton P, Lein J (2000) Eliminating the flaws in New England's reserve markets. Working Paper, University of Maryland, March. www.stoft.com

Cross Border Electricity Trading and Market Design (2004) The England–France interconnector. Business School, London. Reference CS-04-008

Crystal J (1995) Oil and politics in the gulf. Cambridge University Press, New York

Cuevas F (1996) La déréglementation de l'industrie électrique en Amérique Latine: Un premier bilan. Revue de l'Energie, no 480

Daguzen JF (2005) Quel avenir pour un demi-partenariat ? Fondation pour la recherche stratégique. Barcelone, December 1

de Almeida ELF, Pinto H Jr (1999) The driving forces in Brazilian electricity industry. Energy Stud Rev 9(2), Toronto

de Oliveira A (1999) The New Brazilian electricity market. Coordination or Competition, Web Site of the Institute of the Americas, November

Derrieux R (1992) Problèmes et perspectives du secteur électrique dans les pays du bassin méditerranéen: le cas de l'Italie et du Maghreb, Revue de l'énergie, no 439

Dias Leite A (1997) A Energia do Brasil. Nova Fronteira, Rio de Janeiro

Domah P, Pollitt MG, Stern J (2002) Modelling the costs of electricity regulation: evidence of human resource constraints in developing countries

Dubois U (2002) Le modèle allemand de réforme électrique. Evaluation de la régulation par autorité de concurrence. Groupe Réseaux Jean Monet, ADIS et ATOM

Dumez H, Jeunemaitre A (1991) La concurrence en Europe: de nouvelles règles du jeu pour les entreprises. Seuil, Paris

Dumez H, Jeunemaitre A (1998) Evaluer l'action publique: régulation des marchés financers et modèle du mandat. Harmattan, Paris

Dumez H, Jeunemaitre A (2000) Understanding and regulating the market at a time of globalization, the case of the cement industry. Palgrave, New York

Dyllick-Brenzinger RM, Finger M (2013) Review of electricity sector reform in five large, oil- and gas-exporting MENA countries: current status and outlook. Energ Strat Rev 2(1):31–45

Eberhard A, Gratwick K (2005) The Egyptian IPP experience. Management Program Infrastructure, Reform and Regulation. Cape Town, RSA: University of Cape Town Graduate School of Business Breakwater Campus

Eickelman FD (1985) From theocracy to monarchy: authority and legitimacy in inner Oman, 1935–1957. Int J Middle East Stud 17(1):3–24, February

El Katiri L (2014) A roadmap for renewable energy in MENA. OIES, Oxford, January

Elbaradei M (2005) Nuclear non-proliferation and arms control: are we making progress? Carnegie International Non-Proliferation Conference Washington. International Atomic Energy Agency, November 7

Ellig J, Kalt J (orgs.) (1996) New Horizons in natural gas deregulation. Praeger Publishers, London. p 260

Elliot D (2007) Middle East broadens its investments with oil profits. The Wall Street Journal, December 13

El-Sayed Selim M (2001) Arab perceptions of the Euro-Mediterranean Partnership. The Emirates Center for Strategic Studies and Research, the Emirates occasional Papers, No 42

Estrada J, Moe A, Martinsen K (1995) The development of European gas markets: environmental, economic and political perspectives. Wiley, Sussex (England), p 375

Faruqui A, Earle R, Azzouni A (2005) Reforming electricity pricing in the Middle East. MEES XLVIII(49)

Fattouh B, Sen A, Moerenhout T (2016) Striking the right balance? GCC energy pricing reforms in a low price environment. Oxford Institute for Energy Studies, Oxford

Finon D (1993) « Les formes intermédiaires de concurrence dans les industries électriques », les variantes institutionnelles pour les sociétés d'énergie: le cas du secteur électrique. Institut de l'Energie des pays en commun (IEPF)

Finon D (1997) « La concurrence dans les industries électriques: l'efficacité au prix de la complexité transactionnelle et réglementaire? » Economies et sociétés no 7(5–6), pp 13–46

Finon D (2001) Reforms in the French power system: from weak contestability to effective competition? Energy Policy 29(10):749–832

Finon D (2004) Incentives to invest in liberalized electricity industries in the North and South. Differences in the need for suitable institutional arrangements. Working Paper CIRED, Energy Policy, May

Finon D, Kouo D (dir.) (1998) Investissements et nouveaux aspects contractuels dans les industries électriques africaines, IEPF, Quebec

Finon D et al (1996) « La dynamique d'organisation des industries de réseaux énergétiques: l'apport des approches de la nouvelle économie institutionnelle », Cahier de recherche no 7, IEPE, Grenoble

Fitoussi JP (2003) EDF, le marché et l'Europe, Fayard, Paris
Fremaux P (2002) Etats: un retrait en demi-teinte. Alternatives Economiques, no 51, pp 30–31, Trimestre 1
Gilbert JR, Kahn EP (1996) International comparisons of electricity regulation. Cambridge University Press, Cambridge, p 500
Glachant JM (2002) The economics of contracts: theories and applications. Cambridge University Press, Cambridge
Glachant JM (2003a) Quatre designs de réformé électrique. Grande Bretagne: Economie & Sociétés. Série Economie de l'Energie, forthcoming
Glachant JM (2003b) Four sisters weeping on ideal electricity reform. Second Milan European Economy Workshop, June 6/7
Glachant JM (2004) Quatre designs de reforme électrique: Grande Bretagne, Californie, Scandinavie et Allemagne. ADIS, Université de Paris XI
Glachant JM, Pignon V (2002) Nordic electricity congestion's arrangement as a model for Europe: physical constraints or operators' opportunism? Massachusetts Institute of Technology, Cambridge, September 29
Glachant JM, Dubois U, Perez Y (2008) Deregulating with no regulator: is Germany electricity transmission regime institutionally correct? Energy Policy 36(5):1600–1610
Granadino R, Mansouri F (2007) A bridge between two continents. Transmission and Distribution World, May 1
Griffin JM, Puller L (eds) (2005) Electricity deregulation: choices and challenges. University of Chicago Press, Chicago
Guasch JL, Spiller P (1999) Managing the regulatory process: design, concepts, issues, and the Latin America and Caribbean story (English), World Bank Latin American and Caribbean studies; viewpoints. The World Bank, Washington, DC
Haggard S (2000) Interests, institutions, and policy reform. In: Krueger A (ed) Economic reform the second stage. University of Chicago Press, Chicago, pp 21–60
Haggarty L, Shirley M (2002) Telecommunication reform in African countries. World Bank Policy Research Working Papers
Handoussa H, Reiffers JL (2003) L'impact de l'élargissement de l'UE sur les partenaires Méditerranéens. Réseau Femise
Harders C, Legrenzi M (eds) (2008) Beyond regionalism? Regional cooperation, regionalism and regionalization in the Middle East. Ashgate, Surrey
Hausman WJ, Herner P, Wilkins M (2008) Global electrification: multinational enterprise and international finance in the history of the light and power. Cambridge University Press, New York
Head K, Mayer T (2001) Effet Frontière, Intégration économique et « Forteresse Europe ». CEP II, September
Henry C, Matheu M, Jeunemaitre A (eds) (2001) Regulation of network utilities: the European experience. Oxford University Press, Oxford
Hibou B (1998) Economie politique du discours de la Banque Mondiale en Afrique sub-saharienne, du catéchisme économique au fait (et méfait) missionnaire, les Etudes du CERI, no 39, March
Hirschman AO (1958) The strategy of economic development. University Press, Yale
Holburn GLF, Spiller P (2002) Institutional or structural reforms; lessons from international electricity sector reforms. In: Brousseau E, Glachant J-M (eds) The economics of contracts: theories and applications. Cambridge University Press, Cambridge
Hunt S (2002) Making competition work in electricity. Wiley, New York
Hunt S, Shutelworth G (1996) Competition and choice in electricity. Wiley, West Sussex
IAEE (2008) The future of electricity: papers in honor of David Newbery. Energy J
International Bechtel Incorporated (1995) Study to develop a master plan for the power & water requirements during the period 1995–2010. For Government of Abu Dhabi, Water & Electricity Department

Jaime M, Eduardo L, Alejandro M (2001) Sustainability of the electricity sector reform. Inter-American Development Bank, Washington, DC

Jamasb T (2002) Between the state and market, electricity sector reform in developing countries. Util Policy 14(1):14–30

Jamasb T et al (2004) Strategic behavior under regulatory benchmarking. Energy Econ 26 (5):825–843

Jamison M (1999) Pricing and industry structure: the New Rivalry in infrastructure. Kluwer Academic Publishers, Dordrecht

Jancovici JM (2006) Energies, tirer profit de la pénurie. Enjeux hors série, December

Johnson T (2007) The return of resource nationalism. Council of Foreign Relations

Joskow PL (1985a) Vertical integration and long term contracts: the case of coal burning electric generating plants. J Law Econ Org 1:33–80

Joskow PL (1985b) Markets for power: an analysis of electric utility deregulation. The MIT Press, Cambridge

Joskow PL (1987) Contract duration and relationship-specific investments: empirical evidence from coal markets. AER 77(1):168–185, March

Joskow PL (1989) Regulatory failure, regulatory reform, and structural change in the electric power industry. Brookings Papers on Economic Activity

Joskow PL (1991) The role of transaction cost economics in antitrust and public utility regulatory policies. J Law Econ Org 7:53–83

Joskow PL (1996) From hierarchies to market: introducing competition into regulate network industries. Ind Corp Chang 5(2)

Joskow PL (1997) Restructuring, competition and regulatory reform in the U.S. electricity sector. J Econ Perspect 11:119

Joskow P (1998) Electricity sectors in transitions. The Energy Journal 19(2)

Joskow PL (2000) Deregulation and regulatory reform in the U.S. electric power sector. In: Peltzman S, Winston C (eds) Deregulation of network industries: the next steps. Brookings Press, Washington

Joskow P (2001) California's electricity crisis. National Bureau of Economic Research (NBER), Cambridge. Working paper series, no 8442

Joskow P (2002) Electricity sector restructuring and competition: a transactions-cost perspective. In: Brousseau E, Glachant J-M (eds) Economics of contracts: theories and applications. Cambridge University Press, Cambridge

Joskow PL (2006) Markets for power in the U.S.: an interim assessment. Energy J 27(1):1–36

Joskow PL, Schmalsensee E (1983) Markets for power: an analysis of electric utility deregulation. The MIT Press, Cambridge

Joskow P, Tirole J (2003) Merchant transmission investment

Joskow PL, Rose NL, Wolfram CD (1994) Political constraints on executive compensation: evidence from the electric utility industry. NBER Working Paper No w4980, December

Kahn A (1988) The economics of regulation: principles and institutions. MIT Press, Cambridge

Keramane A (1996a) L'énergie et sa distribution. Edisud, Paris

Keramane A (1996b) La privatisation en Asie, Europe et Amérique Latine. OCDE, Paris

Keramane A (1999) Le financement des infrastructures. Revue d'économie financière, no 54

Khatib H (2005) MENA electrical power sector: challenges and opportunities. Middle East Economic Survey XLVIII(39)

King & Spalding LLP (2017) Middle EAST Vision 2030 PPP Legal Report 2017. An overview of PPP legislative frameworks and activities throughout the Middle East

Kleit AN (2006) Electric choices: deregulation and the future of electric power. Rowman & Littlefield Publishers, Lanham

Labaronne D (2001) Privatisation et croissance dans les pays de l'Est. Economie Internationale, la Revue du CPII, no 86, Trimestre 2

Laffont JJ (1996a) Industrial policy and politics. Int J Ind Organ 14:1–27

Laffont JJ (1996b) Privatisations et incitations. Revue économique no 6

Laffont JJ, Maskin E (1982) The theory of incentives: an overview. In: Hildenbrand W (ed) Advences in economic theory. University press, Cambridge
Laffont JJ, Tirole J (1986) Using cost observation to regulate firms. J Polit Econ 94(3):614–641
Laffont JJ, Tirole J (1993) A theory of incentives in procurment and regulation. MIT Press, Cambridge
Leveque F (2006) Competitive electricity markets and sustainability. Cheltenham and Northampton
Levy B, Spiller PT (1994) Institutional foundations of regulatory commitment: a comparative analysis of telecommunications regulation. J Law Econ Org 10(2):201–246
Levy B, Spiller PT (eds) (1996) Regulations, institutions and commitment: comparative studies of telecommunications. Cambridge University Press, Cambridge
Littlechild SC (1998) Some alternative approaches to utility regulation. Econ Aff 28(3):32–37
Littlechild S (2006) Competition and regulation in the UK electricity market. Public Econ 14:3–14
Littlechild SC, Beesley ME (1989) The regulation of privatized monopolies in the United Kingdom. Rand J Econ 20(3):454–472
Lovei L (2000) The single-buyer model: a dangerous path towards competitive electricity markets. Public Policy for the Private Sector, World Bank (llovei@worldbank.org)
Lyon TP (2007) The political economy of regulation. Cheltenham and Northampton
Malik AS, Al Zubeidi S (2003) Electric tariffs based on long-run marginal costs for central grid system of Oman
Marcel V, Mitchell JV (2006) Oil titans: national oil companies in the Middle East. Brookings Institution Press, Baltimore
Maree K (2009) Why developers face higher taxes. MEED, July
Martin JM (1966) Industrialisation et développement énergétique du Brésil. Université de Paris
McCubbins MD, Noll RG, Weingast BR (1987) Administrative procedures as instruments of political control. J Law Econ Org 3(2):243–277
McCubbins MD, Noll RG, Weingast BR (1989) Structure and process, politics and policy: administrative arrangements. Va Law Rev 75:431–482
Mcdaniel T (2000) Utility regulation lectures. GB, University of Cambridge, Department of Applied Economics
Mckinsey S (2007) Reducing US greenhouse gas emissions: how much at what cost. ca
MEED (2009) Aluminum Gulf takes advantage of low costs, July 3–9
Meed Yearbook, First Energy Bank (2008) The world's economic powerhouse
Mejean B (2002) Public-private partnerships for infrastructure development: an international lender's perspective. Barka Power and Desalination project in Oman
Menard C, Shirley M (2000) Reforming public utilities: lessons from urban water systems in six developing countries. University Paris I: ATOM, Working paper
Merlin A (2006) Interview: « l'Europe de l'électricité est déjà une réalité ». Les Echos, September 4
Mestrallet G (2003) Interview. Les Echos, jeudi 9 octobre
Metcalf GE (2008) An empirical analysis of energy intensity and its determinants at the state level. Energy J 29(3). IAEE
Middle East Economic Survey (MEES). (2009) Recent Gulf project financings reveals new price band. A weekly Review of Regional Energy, Finance and Politics LII(33), August 17
Milan J, Vives A (2001) Reform in small electricity markets. Inf Financ Mark Rev 7(2):4–6
Mitsubishi Securities, Clean Energy Finance Committee (2003) Zafarana Wind Power Plant Project. Arab Republic of Egypt: Project Design Document
Moroney JR, Dieck-Assad F (2005) Energy and sustainable development in Mexico. Texas A&M Press, College Station
Nagayama H (2009) Electric power sector reform liberalization models and electric power prices in developing countries: an empirical analysis using international panel data. Energy Econ:463–472
NERA (1997) Proposals for the electricity supply industry in Ireland: consultation paper. Department of transport, energy and communications: an ROINN iompair, Fuinnimh agus Cumarsaide, May

Newbery D (1998) Competition, contracts, and entry in the electricity spot market. RAND J Econ 29(4):726–749

Newbery D (1999) Privatization, restructuring and regulation of network utilities, vol 1, 1st edn. MIT Press, Cambridge

Newbery D (2002) Regulating unbundled network utilities, the economic and social review. Econ Soc Stud 33(1):23–41

Newbery D, Pollitt M (1997) The Restructuring and privatization of Britain's CEGB – was it worth it? J Ind Econ 45(3):269–303

North DC (1989) A transaction cost approach to the historical development of polities and economies. JITE 145(4)

North DC (1990) Institutions, institutional change and economic performance. University Press, Cambridge

North DC (1991) Inst J Econ Perspect 5, Winter

North DC (1993) Institutions and credible commitment. JTE 149

North DC (1994) Privatization, incentives and economic performance. Washington University. Working paper

North DC (1995) Institutions matter. In: Economic history. University Library of Munich, Munich

North DC (2000) Big-bang transformation of economic systems: an introductory note. JTE(156)

North DC, Weingast BR (1989) Constitutions and commitment: the revolution of institutions governing public choice in the 17th century England. J Econ Hist 49:803–832

Observatoire Mediteraneen De L'energie (OME) (2003) Financial Energy Projects in the Southern and Easter Mediterranean Countries

Oliveira A, Pinto JH (1995) La restructuration des industries électriques en Amérique Latine. Revue de l'Energie, no 465

Olson M (1978) Logique de l'action collective. Presses Universitaires de France

Olson M (1982) The rise and decline of nations. Yale University, New Haven

Olson M (2000) Power & prosperity. Basic Books, New York

OME (2003) Med supply? Development of energy supplies to Europe from the Southern and Eastern mediteranean countries

Percebois J (1992) Les réseaux électriques dans les pays en développement. Revue de l'énergie, no 438

Percebois J (2008) Electricity liberalization in the European Union: balancing benefits and risks. Energy J 29(1), IAEE

Percebois J, Wright P (2001) Electricity consumer under the state and the market: French and U.K. electricity supply industries. Util Policy 10(3):167–179

Perez Y (2002) l'analyse néo-institutionnelle des réformes électriques européennes. Thèse pour le doctorat en sc. Economiques. Université de Paris I Panthéon- Sorbonne

Pinto JH (1993) Financement, Investissement et Modes d'Organisation des Industries Electriques: le cas des pays d'Amerique Latine. Doctorate Thesis at Institute of Energy Policy and Economics/Grenoble II University

Pinto JH (2002) Institutional designs and regulatory reforms in the energy industries: an international comparative analysis and lessons for Brazil, In Oxford-BP Brasil Research Paper no 1, University of Oxford, June

PLATT'S (2005) Euro-Maghreb power export projects have political backing, April

Qadous M (2008) Determining of electricity tariff in Jordan, Director of Studies and Economic Researches Department, ERC Jordan, Tariff/Pricing Committee Meeting, Budva, Montenegro, February 4–5

Quast O (1997) Les Fondements des Modèles Successifs d'Organisation de l'Industrie Gazière: Analyse des Conditions d'Introduction de la Concurrence dans une Industrie de Réseau. Phd Thesis at Institut d'Economie et de Politique de l'Energie. Grenoble, France, p 308

Rapin D, Vassilopoulos P (2004a) L'évolution des prix de l'électricité en univers incertain: Contractualisation et gestion du risque prix, Revue de l'Energie, September

Rapin D, Vassilopoulos P (2004b, Juillet-Août), Ouverture à la concurrence du secteur électrique en Europe: Ce qui va évoluer, Revue Accomex

Razavi H (2009) Natural gas pricing in countries of the Middle East and North Africa. Energy J 30 (3), IAEE

Rees R (2006) The economics of public utilities. Edouard Elgar, Cheltenham and Northampton

Roques F, Newberry D, Nuttall W et al (2005a) Valuing portfolio diversification for a utility: application to a nuclear power investment when fuel, electricity and carbon prices are uncertain. Cambridge

Roques FA, Newberry DM, Nuttall WJ (2005b) Marchés électriques, Sécurité Energétique et Diversification Technologique: le Nucléaire comme Couverture face aux risques de Prix du Gaz et du Carbone? University of Cambridge, Angleterre, août 12

Roques FA, Nuttal WJ, Newberry DM et al (2006) Nuclear power: a hedge against uncertain gas and carbon prices? Energy J 27(4), IAEE

Ruet J (2003) Optimal timing in privatization of a utility in an emerging country. In: The case of electricity distrbution in Delhi. CERNA, Paris

Ruet J (2005) Privatizing power cuts? In: Ownership and reform of state electricity boards in India. CERNA, Paris

Ruffin C (2002) The political economy of institutional change in the electricity supply industry: shifting currents. Edward Elgar Publishing, Cheltenham

Sadiki L (1997) Towards Arab liberal governance: from the democracy of bread to the democracy of the vote. Third World Q 18(1):127–148, March

Salazar Guerra M (2003) Mexico's electricity reforms: effects of private electricity generation and their role to enhance a sustainable electricity industry: Master Thesis. Lund University, Department of Energy and the Environment, November

Saudi University (2004) Electric energy production costing for the Saudi electricity sector (phase-1). Prepared for Electricity and Co-generation Regulatory Authority (ECRA), Riyad

Saussier S (1997) Choix contractuels et coûts de transaction: une analyse économique des contrats d'approvisionnement en charbon d'EDF. Thèse de doctorat, Université ParisI Panthéon Sobonne

Sautter C (1973) Japon. le prix de la puissance. Seuil, Paris

Savedoff WD, Spiller PT (1999) Reforms of water distribution

Sen A, Nepal R, Jamasb T (2016) Reforming electricity reforms? Empirical evidence from Asian economies. Oxford Institute for Energy Studies, Oxford

Shi X, Polenske KR (2006) Energy prices and energy intensity in China. A structural decomposition analysis and econometrics study. Massachusetts Institute of Technology, Center for Energy and Environmental Policy Research (Cambridge). Working paper number 606, May

Sioshansi FP (2008) Competitive electricity markets: design, implementation, performance. Elsevier, Amsterdam

Sioshansi FP, Pfaffenberger W (2006) Electricity market reform: an international perspective. Elsevier, Amsterdam

Smith C (2004) Az Zabirah Aluminum Project. Dubai

Spiller PT (1993) Institutions and regulatory commitment in utilities' privatization. Ind Corp Chang 2(3)

Spiller PT (1996a) A positive political theory of regulatory instruments: contracts, administrative law or regulatory specificity? Southern California law review 69:477–515

Spiller PT (1996b) Institutions and commitment. Ind Corp Chang 5(2)

Spiller PT, Holburn GLF (2001) Institutional or structural: lessons from international electricity sector reforms. University of California Berkeley. Working paper, September

Spurr M (2005) District cooling and Middle East infrastructure policy

Stark P (2007) The winds of change: resource nationalism shifts the balance of power to national oil companies. JPT, Society of Petroleum Engineers 59(1), January

Steiner F (2000) Regulation, industry structure and performance in the electricity supply industry. OECD, Economics department Working Papers, Paris. p 238

Stern J (1998) Competition and liberalisation in European gas markets: a diversity of models. The Royal Institute of International Affairs, London, p 232
Stern J, Holder S (1999) Regulatory governance: criteria for assessing the performance of regulatory systems. An application to infrastructure industries in the developing countries of Asia. Util Policy 8:33–50
Stiglitz JE (1987) Principal and agent. Dans Eatwell J (ed), The New Palgrave
Stiglitz JE (1998) More instruments and broader goals: moving toward the post-Washington consensus. The World Bank (Helsinki): The 1998 WIDER Annual Lecture, pp 6–9
Stiglitz JE (1999) Whither reform? Ten years of the transition. World Bank, Annual Bank Conference on Development Economics, p 16
Stiglitz JE (2009) Progress? What progress?. OECD Observer no 272, March
Surrey J (ed) (1996) The British electricity experiment privatization: the record, the issues, the lessons. Earthscan, London
Tenenbaum N et al (1992) Electricity privatization. Structural, competitive and regulatory options. Energy policy, December
Teplitz-Sembitzky W (1990) Regulation, deregulation, or reregulation – what is needed in the LDCs power sector. The World Bank Industry and Energy Department, Energy Series Paper, no 30
The European Commission (EuropeAid Cooperation Office DG) (2003) Ministerial declaration of the Euro-Mediterranean energy forum: Euro-Mediterranean ministerial conference on energy. Athens, May 23
The World Bank Group, the Energy and Mining Sector Board (2004) Operational guidance for World Bank group staff
The World Trade Organization (2005) Report of the working party on the accession of the Kingdom of Saudi Arabia to the World Trade Organization. WT/ACC/SAU/61, November
Thomas Y (1994) Les producteurs indépendants d'électricité en Europe. InterEditions, Paris
Tsebelis G (1995) Decision making in political systems: veto players in presidentialism. Br J Polit Sci 25:289–325
US-Saudi Arabian Business Council (2005) Power generation and distribution in the Kingdom of Saudi Arabia, December
Varoquaux W (1996) Calcul économique et électricité. PUF, Paris
Vassilopoulos P (2003) Models for the identification of market power in wholesale electricity markets. DEA 129, Industrial Organization. Université Paris IX, Dauphine. U.F.R. Sciences of Organizations, September
Victor DG, Heller TC (2007) The political economy of power sector reform: the experiences of five major developing countries. Cambridge University Press, New York
Villain MJP (2002) First advisor, investment strategy, Abu Dhabi Investment Authority (ADIA), May 22
Walker I, Benavides JM (2002) Sustainability of power sector reform in Latin America. In: The reform in honduras. Working Paper. Inter-American Development Bank, Washington
Wallsten SJ (2001a) An econometric analysis of telecom competition, privatization, and regulation in Africa and Latin America. J Ind Econ 49(1):1–19
Wallsten S (2001b) Ringing in the 20th century: the effects of state monopolies, private ownership and operating licenses on telecommunications in Europe, 1892–1914 Stanford Institute for Economic Policy Research, Working Paper
Wallsten S (2002) Does sequencing matter? Regulation & privatisation in telecommunication reform, World Bank working paper no. 2817
Weinmann J, Bunn D (2004) Resource endowment and electricity sector reform
Williamson O (1975) Markets and Herarchies: analysis and anti-trust implications. Free Press, New York
Williamson OE (1976) Franchise bidding for natural monopolies- In general and with respect to CATV. Bell J Econ 7(1), ping
Williamson OE (1979) Transaction-cost economics: the governance of contractual relations. J Law Econ 22(2), Octember

Williamson OE (1983) Credible commitments: using hostage to support exchange. AER 73(4), September
Williamson OE (1985) The economic insitutions of capitalism. Free Press, New York
Williamson OE (1989) Transaction cost economics. In: Schmalensee RL, Willig RD (eds) Handbook of industrial organization. North Holland, Amsterdam
Williamson OE (1991) Comparative economic organization: the analysis of discrete structural alternatives. Adm Sci Q 36(2)
Williamson OE (1993a) Transaction cost economics and organization theory. Ind Corp Chang 2(2)
Williamson OE (1993b) Transaction cost economics: the governance of contractual relations. J Law Econ 22(2)
Williamson OE (1995) Hierarquies, market and power in the economy; an economic perspective. Ind Corp Chang 4(1)
Williamson OE (1996) The institutions and governance of economic development and reform. In: Williamson OE (ed) The mechanisms of governance. Oxford University Press, New York
Williamson OE (2002) The theory of the firm as governance structure: from choice to contract. J Econ Perspect 16(3)
Williamson OE, Masten S (1999) The economics of transaction costs. Elgar critical writings reader
Wolak FA (1997) Market design and the behavior of prices in restructured electricity markets: an international comparison. Stanford University, Working Paper
Wolak FA, Patrick RH (1997) The impact of market rules and market structure on the price determination process in the England and Wales electricity market. University of California, Energy Institute Power Working, Paper PWP-047
Wolak FA, Patrick RH (2001) The impact market rules and market structure on the price determination process in the England and Wales electricity market. NBER Working Paper No w8248, April
Wolak FA, Nordhaus R, Shapiro C (1998) Preliminary report on the operation of the ancillary services markets of the California independent system operator (ISO). Folsom, Calif, California Independent System Operator
Wolfram C (1999) Measuring duopoly power in the British electricity spot market. Am Econ Rev 89:805
Wolfram C (2000) Electricity markets: should the rest of the world adopt the United Kingdom's reforms? Regulation 22:48
World Bank (1995) Bureaucrats in business: the economics and politics of government ownership. Oxford University Press, Oxford, p 346
World Bank (1997) Expanding the measure of wealth: indicators of environmentally sustainable development. World Bank, Washington
World Bank (2002) Building institutions for markets. The World Bank, Washington
World Bank (2006) Where is the wealth of the nations? The World Bank, Washington
World Bank Hamilton K (2005) Testing Genuine Saving. World Bank Policy research Working Paper No 3577
Wright V (1993) Les privatisations en Europe, programmes et problèmes. Actes Sud, Poitiers
Yergin D (2009) Ensuring energy security. Foreign Aff 85(2)
Yergin D, Markovich L, Stauffer H (2003) The problem with power: the system did not fail. Yet the system failed. The New York Times, August 17
Zelner B, Henisz W (2000) Political constraints, interest group competition and infrastructure investment in the electricity utility industry: a cross national study, Working Paper of the Reginald H. Jones Center, Philadelphia
Zhang Y, Kirkpatrick C (2002) Electric sector reforms in developing countries: an economic assessment of the effects of privatization, competition and regulation. Center on Regulation and Competition, Working Paper No 31, University of Manchester

Index

L. Benali, *Electricity-sector Reforms in the MENA Region*, Perspectives on Development in the Middle East and North Africa (MENA) Region,
https://doi.org/10.1007/978-3-319-96268-9

The manufacturer's authorised representative in the EU is Springer Nature Customer Service Centre GmbH, Europaplatz 3, 69115 Heidelberg, Germany. If you have any concerns regarding our products, please contact ProductSafety@springernature.com

Printed and bound by CPI Group (UK) Ltd, Croydon, CR0 4YY

30/09/2026

02225662-0004